ADVANCE PRAISE (Alphabetic

"For anyone interested in building online communities, Community 101 *is a must-read guide. Robyn and Miranda have written a truly helpful resource for people who are trying to grow meaningful relationships with their virtual communities. The authors have a passion for approaching online communities with authenticity, respect, and empathy, applying the Golden Rule time after time. I look forward to re-reading* Community 101 *many times over as I strive to build lasting communities both online and offline."*

Julie Shin Choi, Entrepreneur and Community Builder

"Building an online community is a two-part challenge: 1) knowing the rules, and 2) applying them at internet (translation: warp) speed. In simple terms, Robyn and Miranda lay out all the rules and tell you exactly what to do with them. Every community manager's desk should have a dog-eared, heavily highlighted copy of this clear-cut primer."

Easton Ellsworth, President and Co-Founder, SocialKen

"I am a vocal proponent that everything online is an extension of what people do offline; only the tools are different. This book converts everything we know about offline communities and provides a practical, step-by-step guide to building a strong, effective online community. Three thumbs up!"

David Leonhardt, President, The Happy Guy Marketing

"Finding and connecting with the right people is part of the social glue that hold personal and professional relationships together on the Internet. Robyn and Miranda go through a detailed process of identifying, establishing and building an online presence by combining various strategies and tools to enhance your success."

Andrew Wee, Blogger and Business Consultant
WhoIsAndrewWee.com

Community 101

How to Grow an Online Community

By Robyn Tippins
and Miranda Marquit

Foreword by Dave Taylor

20660 Stevens Creek Blvd., Suite 210
Cupertino, CA 95014

Published by Happy About®
20660 Stevens Creek Blvd., Suite 210, Cupertino, CA 95014
http://happyabout.com

First Printing: October 2010
Paperback ISBN: 978-1-60005-152-4 (1-60005-152-9)
eBook ISBN: 978-1-60005-153-1 (1-60005-153-7)
Place of Publication: Silicon Valley, California, USA
Paperback Library of Congress Number: 2009942922

Trademarks

Warning and Disclaimer

Dedication

To my mom, who always believes I'm far smarter than I am. Without her support, I'd have never imagined I could do this.

—**Robyn**

To Josh and Gavin. My boys.

—**Miranda**

A Message from Happy About®

Thank you for your purchase of this Happy About book. It is available online at http://happyabout.com/community101.php or at other online and physical bookstores.

- Please contact us for quantity discounts at sales@happyabout.info
- If you want to be informed by email of upcoming Happy About® books, please email bookupdate@happyabout.info

Happy About is interested in you if you are an author who would like to submit a non-fiction book proposal or a corporation that would like to have a book written for you. Please contact us by email editorial@happyabout.info or phone (1-408-257-3000).

Other Happy About books available include:

- 18 Rules of Community Engagement: http://www.happyabout.info/community-engagement.php
- Social Media Success!: http://www.happyabout.com/social-media-success.php
- I Need a Killer Press Release—Now What???: http://happyabout.info/killer-press-release.php
- I'm on LinkedIn—Now What???: http://happyabout.info/linkedinhelp.php
- I'm on Facebook—Now What???: http://happyabout.info/facebook.php
- 42 Rules for Successful Collaboration: http://www.happyabout.info/42rules/successful-collaboration.php
- 42 Rules for Effective Connections: http://happyabout.info/42rules/effectiveconnections.php
- 42 Rules for 24-Hour Success on LinkedIn: http://happyabout.info/42rules/24hr-success-linkedin.php
- 42 Rules™ to Jumpstart Your Professional Success: http://happyabout.info/42rules/jumpstartprofessionalservices.php
- 42 Rules of Social Media for Small Business: http://happyabout.info/42rules/social-media-business.php
- Rule #1: Stop Talking!: http://happyabout.info/listenerspress/stoptalking.php
- The Successful Introvert: http://happyabout.info/thesuccessfulintrovert.php

Contents

Foreword by Dave Taylor

Robyn starts this book by talking about a community she helped created in 1998. I can remember a decade prior to that the virtual community my friends—old and new—had created through something called Usenet.

Usenet wasn't glamorous and at that point our computer systems displayed only text. No graphics, no buttons, no fancy windows, just a screen of text and a set of complex commands you typed in to work with the system. And yet I remember the friends I made, some of whom I still count as friends today.

That's because a community, in reality, is about people and how those people interact with each other and learn to appreciate each other's similarities and differences, and also about how we all learn to empathize with and care about one another.

Robyn and Miranda know very well that a community is all about people, and that the secrets to building a healthy online community revolve around how you interact with the people in that community, not the technology, color schemes, skins or typefaces available—and that's what this book is all about.

In this book the authors talk about ROI, return on investment, for companies creating communities, but I suggest that you shouldn't quantify it in terms of money, customer retention or sales, but instead focus on two basic factors: market segment visibility and customer happiness. Happy customers are evangelists, as Apple Computer demonstrates quarter after quarter.

The authors also offer five basic rules for creating a successful online community: tell it like it is, work towards positive change, be visible and active, keep tweaking and tuning the site and community guidelines and remember the 'do unto others' golden rule. They're spot on...because the rules of creating a successful virtual community are the same as for any other kind of community: listen along with your talking, and be honest, engaging and polite.

Now, implementing these rules consistently and encouraging your community members to embody them too? That's not so easy after all, and that's exactly why you need to say 'enough foreword, let's get on with it!'

And so I'll wrap up with my golden rule: stop writing when you've made your point! :-)

Enjoy the rest of this book and I'm sure I'll see you online in one community or another.

Dave Taylor
Boulder, Colorado

Dave Taylor, author of film reviews at DaveOnFilm.com, is a member of more virtual communities than he can count. Operator of the popular site AskDaveTaylor.com, he counts his friends by the thousands. By the time you read this book, Dave will be celebrating thirty years online.

It's difficult to believe, but a debate exists about whether or not online, or 'virtual,' communities are 'real.' To delve into that discussion, search for 'virtual community' on Wikipedia,[1] and you'll see that just like every topic under the sun, people will debate just about anything online. On this discussion in particular, many do not think there are strong enough ties in most virtual communities to recognize them as real communities. There are others, though, who disagree and insist that virtual communities are every bit as close-knit and cohesive as real-life communities.

Robyn says:

> "I've been involved in virtual communities since 1995, and have formed some very strong connections to the men and women I've met over the years. These are the people I ask when I need advice on how to deal with work and family, and, thanks to their support and advice, I'm a better person. Some of the communities I've interacted with over the years exist solely online, and many of these have became so important to real life that some of the members began to meet offline, even when we had to travel across several state lines to do so.
>
> One of my favorite early online communities was started to chronicle the months of my first pregnancy. The 'FebMoms' club, on iVillage, was a part of an online women's

---

1. Wikipedia: "Virtual Community," http://en.wikipedia.org/wiki/Virtual_community

community back in 1998. I interacted with roughly 30 moms on a daily basis, usually via email, but often in the online forums as well. We shared the daily joys and pains of pregnancy. From intimate discourses on body changes to how we were going to decorate our nurseries, we lived our lives together, and the women became my best friends. Our camaraderie was based on this shared experience, and it would have been nearly impossible to find, locally, 30 women who all shared the same birth month. This community could only be found online.

We saw several members drop out when their pregnancies ended both naturally, or with medical help. We supported each other through miscarriages, abortions, or still-births, and rejoiced at each healthy birth. Our group continued into the first year of our babies' lives, commiserating the trouble regaining our pre-baby bodies; discussing the difficulties of post-partum sex; railing against the issues we each had with ob-gyns, pediatricians, labor coaches and daycare workers; arguing the perpetual debate of breast vs. bottle and working moms vs. stay-at-home-moms, and generally we both hated and enjoyed each other's company. For me, no offline community could have given me more enjoyment and more frustration. These moms were as much a community as my church, my extended family and my neighbors. I both hated and loved these moms, every bit as I dually hate and love the offline communities I enjoy. Real community? I think so!"

Regardless of your perspective on this debate, as a company, in a world which increasingly does business online, and where customers can be found all over the world via the web, learning to build an online community becomes as vital to your bottom line as optimizing your website for search engines or ensuring that your PR department gets your messaging out to the right outlets. It's now a necessity in business to have a social strategy, so sit up and pay attention!

The More Things Change...

The more technology changes the way we live and interact, the more things stay the same. In many ways, online communities are similar to the communities of our predecessors. People have always been drawn together by common interests, like geography (local communities) or identity (families, union workers, political groups). It's much the same online. The Huffington Post is built up of a community that revolves around the politics of progressive, democratic Americans (identity).

Craigslist is hyper local, with people from cities across the world, interacting with their own local communities (geography). We still come together based on geography and identity, much the same as we did, way back when.

The way we act when we come together hasn't really changed either. People have always vigorously debated with those who don't agree with them. As before, reputation among the members of the community is key. Debate is still one way of solving problems and winning over those who disagree. Debate also makes us feel superior, more tightly binds members of communities, and solidifies our own opinions.

We base our decisions about these debates on the reputations of those arguing. In offline communities, a debate about county taxes will ensure we give more credence to the County Administrator than we do to the out-of-work actor who lives two states away. But more than just relevance, reputation matters. We trust the opinion of Dr. Oz when the question is medically related, even more than we trust our local doctor, because Dr. Oz's reputation is bigger and better. Likewise, in online communities, we look at the number of posts someone has written, how much everyone else respects or reviles those posts, or the relevance of their expertise to the debate, and we come up with an internal score to judge their level of expertise in the community. Newcomers, online and offline, must prove themselves worthy of our trust.

Online outcasts are as common a commodity as community outcasts in the real world. We stay away from the shady guy who hangs around preschools. We base our offline assumptions of reputation on how people look, talk, act, or smell and even if they appear to have confidence. Robyn likes to say that the world is just like high school, no matter where you are. There are niche groups (jocks, nerds, brains) in every community. The difference on the web is that these groups can shut out the other groups and better carry on their conversations without the same fear of persecution. This is not to say that there aren't horrible people online, just as there are horrible people offline. The same person we call a troll online, is probably a miserable, undesirable person offline as well. And trolls didn't just happen when online communities came about. Trolls were there all those years ago, even though we called them by different names.

Even the way we treat each other is the same. In offline communities, even before the Internet, people spoke more nicely to your face than behind your back. Buttering up a journalist, now and then, was a great way to get favorable coverage for your company. All of that is almost completely unchanged in online communities, except now you're much more likely to get caught gaming the system.

Online communities bring age-old human interactions to the web. The number of sites devoted to specific interests is growing, allowing nearly anyone to find a community to belong to. While there are only a handful of bloggers in our hometown, we can log on and find thousands to discuss the daily woes of being your own boss, supporting yourself merely through advertising with a good dose of effort and talent. Online, people band together for support, human emotional contact and protection—just as they do in real life, but you're far more likely to find carbon-copies of yourself around the globe than to hope that they exist in the limited pool of people in your small town.

Application to Your Business

Reputation represents one of the most important carry-overs from the offline to the online world. Your reputation on the web is always tested. Are you ethical? Are you a responsible member of the community? How long have you been a member? Do you lurk, or do you frequently post? Do people trust you? Are your actions online consistent with your actions offline? Do you offer credible and reliable information?

Just as how your standing in the corporate or political world, or how consumers view you, affects your offline business ventures, your online success depends on how members of your web community view you and your brand.

Debate is a mainstay of community that carries over into cyberspace. We debate in the real world, considering the merits of positions, ideas, products and information. And we love to endlessly discuss things when someone has screwed the pooch.

Almost all of us have been in the position of engaging in lengthy dis-course, of very little real import, simply because someone on the Internet was (gasp!) wrong. Debate is of vast importance online. And it's fun. From discussions on theology to whether Splenda will kill you, members of virtual communities will debate for hours, days even, and usually end up without a conclusion. This debate, though, brings people together. It enriches your community and deepens bonds. As in real life, we fight with people and then we make up. This exchange of raw emotions creates a schism which, when mended, bonds participants more deeply than if they only exchanged pleasantries all day long. We fight with our kids, spouses, siblings and parents. We do not fight with our PTA rep, the cashier at Target, or others who are not truly important to our lives. When you can love someone enough to rail at them, you know that you are truly bonded. Debate makes the heart grow fonder, I suppose.

While this means you can improve your company's position by joining the debate offline, it also means the Internet offers opportunities to spark debate in the virtual world. As you might expect, debate draws people (virality). Consumers pay attention to debates in their communities. They are drawn to new communities when they see an exciting, relevant discussion. This truth gains greater importance and more voice online, where anyone can contribute to the dialogue, and usually does so, both on Facebook and Twitter, and with their blog.

Beware, though: Just as you can make mistakes in the real world that will turn you and your company into a sort of social pariah, it is possible to find yourself labeled an undesirable in the online realm. It's not only possible, but it's also probable, if you don't handle yourself well. The key to avoiding that fate rests in your ability to effectively build and nurture an online community.

New Ways of Forming Communities Mean New Rules

Of course, nothing stays exactly the same, and online communities are no exception. They differ from offline communities in key ways. Geography is less important in web communities. Physical presence is not a requirement for community participation. Geographic identity is much more liberal in an online community.

Geography matters much less than it once did, though with apps like Gowalla and Foursquare, that's changing. Before the widespread popularity of the web, community definition was limited largely by physical location. Your community was your neighborhood, your school and your city. In the old days, it was quite difficult to leave your community, especially in the times before easy and affordable motorized travel. But online, if you want to leave your virtual community, you can simply delete your account. Of course, in some communities this is easier said than done, but it's a conclusion that can be obtained without massive upheaval of your offline life. This is a much easier solution than moving to another geographic area.

Likewise, physical presence is no longer required to participate in a community. You do not have to head to the town square with your soap box if you want to voice your opinion. Just log on to a community discussion board and tell people *exactly* what you think. Going to your monthly book club meeting no longer means you must travel to someone's house, find a sitter and bake a sugary confection. Book discussions online allow you to attend from the comfort of your own home, in robe and pajamas, long after the kids are in bed. Geography is not a barrier.

Identity is its own complex topic. Way back when, your identity was inextricably connected to your family, your co-workers and your friends. Whether your parents were demons or saints, that identity inevitably passed on to you. Was your mom a doctor or the town whore? Did your dad spend years at Oxford or in jail? Were your grandparents respected members of the clergy or wandering carnies? With an online community, you have a more fluid identity that you can change as you grow. Robyn has been widely identified as a mom, music buff, mountain climber, avid romance novel devourer and speaker on social media, and jumps from circle to circle with very little difficulty. Identity can encompass all your complexity and intricacies.

Still, reputation is a factor. In Robyn's example, if those niches were polar opposites. Say she was once roundly opposed to something and is now all for it—she'd have some explaining to do. Because, while you can change and grow, just as you can offline, the web has a way of bringing things up again and you will have to answer for anything you've said or done that doesn't jive with what you are currently saying. And, while you can recreate yourself, if you are fairly well known, this process will be much more difficult. You can have multiple identities, and hide your offline identity from the online world, but as the web matures, this is becoming much more difficult, and that is a good thing.

Online identity is more flexible than the offline identity, but the online identity is becoming more fixed. It's a case where the new rule is actually being replaced by the old rule. As social networks become more important, and as online communities grow in popularity and become more real, your online identity is losing some of its fluidity. Everyone you know online associates you with a specific screen name, and that avatar personifies your identity.

What happens to your reputation when you change your identity? Will you delete all your accounts with that screen name? Will you delete the blog associated with that persona? As that connection from blog, to Twitter ID, to LinkedIn profile, to Facebook account, to OpenID domain, to the myriad other social media profiles you have, gets even more blurred and becomes more of a direct link to the real you, removing yourself will become more complicated. That's why it's important to carefully consider your online identity, and strive for consistency.

Why Online Community Matters

As the web develops, online community will only become more impor-tant. Social networking through online community membership is es-tablished normalcy today, and we don't expect that to change considerably in the near future. This is especially important for busi-nesses to take into consideration. Customers want interaction with companies, and social media allows this to happen, whether on your own site or on any number of third-party sites. Unless your company is an unknown entity, a conversation will happen, involving your company or service. The only real question is whether your own voice will be part of the discussion.

Consider this: Facebook reports over 400 million active users.[2] MySpace boasts well over 113 million monthly active users.[3] Yahoo! and Google are tapping into their own social graphs, with Yahoo! Updates and Google Buzz. Even if you take into account multiple identities and inactive accounts, a large audience can be reached via social networking. Ask the Arctic Monkeys: Without MySpace, they might never have reached such a wide audience. OKGo! may never have had their amazing success without the viral treadmill dances they posted on YouTube. Social networking is paving the way for new opportunities—but only if businesses learn to build community effectively and appropriately.

Social media is the natural evolution to the web, but it is so much more than viral marketing. Companies could feasibly be made—or unmade—based on how well they execute their online community strategy. Web 2.0 represented a huge step in user interaction on the Internet. As the next iteration of the net evolves, social networking is going to become even more important. We're already seeing consumers demand more in terms of business interaction, accountability and ease of use, online. If your company doesn't start building a community now, someone else will, and you may be left out of your own discussion.

---

2. Facebook member statistics,
http://www.facebook.com/press/info.php?statistics.
3. Sarah Tabil and Adam Satariano, "MySpace Appoints Former Facebook Executive as Chief," Bloomberg, http://bit.ly/b21Xz0
(www.bloomberg.com/apps/news?pid=20601103&sid=arxlMc7Zx0zU&refe r=us).

1 What It Means to Build Community

Now we have a solid grasp of what online communities *are*. It's time to tackle what it *means* to build community. In both online and offline communities, it's evident that there are different types of groups, and different approaches to the way leaders in each community behave. There are some communities where the participants are active, and the subject matter is infinite. On the other hand, there are communities where the subject matter is finite, and any off-topic chatter is discouraged. And, there are others where people belong simply for the sake of belonging—think Facebook Fan pages—and no matter the subject matter, they may not take part in debate and may be there simply to digest information. There is overlap among these types of communities as well.

A book club often has a set structure and a finite topic. Members may frown upon discussions about personal matters or religion. In this community, each member is expected to share his or her thoughts about the book, and 'lurking' is discouraged. Another common community is a parent/teacher organization. While the topics may vary, the overall aim remains the same. Lurkers are encouraged to soak up information, though, and there are clear leaders who spend a large amount of their time being active in the group. Topics in both groups change on a regular basis, but the theme of the group never does. Participation may vary, but the fact that participation by some members is always present, does not.

Online communities are similar. You might notice that some websites are fairly static. They post information, but they do not actively support two-way conversations. They may surface relevant pictures, but they do not offer ways to encourage community participation in either uploading their own relevant pictures, or by offering ways to comment on, or even remove, pictures. Other online portals seem to be really alive. They encourage discussion and look for input from community members. In many ways, these variances are similar to the difference between attending a speech and attending a panel discussion, and the resulting effects on community members reflect these dissimilarities.

Speeches are distinctly different from panel discussions that offer a question/answer portion. The speech may be fiery—and even interesting. It may be accompanied by visuals or music. But beyond showing up and maybe doing a little networking afterwards, there's not a great deal for the listener to do. Audience participation is passive. You may take notes, you may even blog about it later, but the speaker will likely never know how you feel about what he's said. This is a classic example of one-way communication.

A panel discussion, on the other hand, requires active participation. There are more players, which means you get the points of view of several people. You can ask questions and share your opinions as part of the audience. You benefit from the questions that other audience members ask. You become part of a discussion that you must actively engage in, even if you don't ask a question yourself. The spotlight is on you and your peers almost as much as it is on the speakers. It can involve a variety of experiences and interactive moments. Networking and education happens in both experiences, but in communities that are alive, the give and take is sometimes so rewarding that it takes both the topic and your participation to the next level.

If you could choose to participate in an activity that featured panel discussions, with Q/A, over an activity, like a speech, that meant you sat in a chair and only digested information, which would you choose? Do you think the people you are trying to reach are any different? People thrive on knowing that they are being heard and that they are offering you, and those around you, value with their input. Are you giving speeches to the members of your community, or are you giving them an opportunity to respond?

If you are not giving them a chance to talk back, your community will gradually shrink as members defect to communities that make efforts to value their input.

Moving Beyond the Static

Not too long ago, web pages were mostly built in a static state. They didn't change very often. You'd post information, and it stayed on your site for months. Attention business owners: the web has evolved! Gone are the days when a static page could suffice as your web presence. In this world, community members demand so much more. Many businesses are listening, investing thousands of dollars to upgrade their static homes on the web, remodeling them into thriving places where community can take root.

If your goal is to increase traffic to your pages, your site must have changed since a user's last visit. There is very little reason for a mother to visit her daughter's soccer team site if there is rarely any new information on it. Likewise, your site is not worth your customer's time if everything on it is outdated and of little value.

An easy way to make sure there is always something of value and interest to your visitors is to use dynamic content. RSS feeds can deliver images, video, text and audio to your site automatically. Internet users expect this. From the beginning, when a customer visits your site, and other similar sites, they expect to find store locations, hours and a page detailing the company. In addition to this basic information, visitors to your site now require information on current sales, the opportunity to make an appointment (if available), coupons, testimonials from other customers, suggestions on how to use your products and services, links to companion sites, tutorials and maybe even previews to upcoming events. People expect your site to be just as important to them and to you as your stores. They expect an online experience that is as relevant as the offline one.

Frequent updates are mandatory. This is a primary reason behind the explosion of blogging as an important community-building tool. Blogs make it easy to update content, as well as add a variety of media to a website. Blogs are readymade for community interaction, providing a way for members to post comments and share their ideas, with or without your constant policing.

Many businesses also integrate chats, online workshops, webinars, and forums to increase interaction on their websites. You can encourage community members to submit their own video tutorials, or create podcasts, based on the product you sell or the industry in which you participate. Part of building a community is encouraging member participation. Consider that companies who do not encourage feedback (both good and bad) are stifling their customers. They are placing a muzzle on what could be complimentary or constructive feedback, and that's never good for business. Additionally, by refusing to listen to negative feedback, companies miss out on the opportunity to learn what they can do to better serve community members. Dan and Chip Heath, authors of Made to Stick, are big fans of the 'customer thank you.' In an article for Fast Company magazine, the authors cite a study—conducted by Sonja Lyubomirsky, a researcher at the University of California Riverside—that looked at strategies to make yourself happier. According to the study, the first was expressing gratitude. The Heath brothers explain that a customer's thanks 'creates a halo of happiness.'[4] Feedback, positive or negative, will help your community members feel as though they are just as invested in the community's success as you are, essentially part of a family. Emotionally invested community members can become both loyal customers and enthusiastic fans.

More than ROI

One of the staples of business thinking for decades has been return on investment (ROI). Businesses want to feel as though they are getting a real bang for their buck, and they should. ROI is never something to ignore, but with community efforts, the way we calculate it must be adjusted. It should come as no surprise that ROI is a key factor that

---

4. Dan Heath and Chip Heath, "Made to Stick," *Fast Company* (October 2008): 95–96.

stakeholders use to sell community building efforts to decision makers in the corporate world. After all, it takes time, effort and money to redesign your website so that it is conducive to community. Spending must also take into account the cost of management of this community, which usually requires new headcount for the day-to-day task of policing and messaging. By far, effort is the most expensive part of community redesigns. The technology is not very expensive—it's often free—but the task of doing the work of community can be extremely time-consuming, especially in the beginning.

ROI is vital in determining the success of marketing efforts, but the ROI on a community site is often difficult to ascertain, especially in the early stages. Building community online requires you to think beyond the bottom line, including how much traffic your community is driving to your business web address. Looking beyond ROI forces you to see your customers as more than just numbers and statistics. You come to see them as community members and potential community members. But numbers matter, and they should always matter. 'We count people because people count' is a mantra we've all heard and it's no less important today with our communities spread across multiple sites and interested in multiple topics. The key is to strike a balance between ruthlessly driving traffic and building a community of members that feels unique and appreciated.

But it's not impossible to measure ROI. Dell has done so with their innovative Twitter marketing. In June of 2009, Dell announced that they had reached over $3M in sales that were directly related to traffic from Twitter.[5] But, just as important as direct revenue increases, is the amount of money you might save if you use online channels to serve your customers. Think how much it costs you to help someone use your product via Facebook, versus the cost of technical help via a call center. Consider the savings of using Twitter to promote a campaign, rather than an expensive PR firm. The cost-saving uses of social media shouldn't be discounted, and we can't only look at making money, but saving it as well, when we measure ROI.

---

5. Marshall Kirkpatrick, "Social Media ROI: Dell's $3m on Twitter and Four Better Examples," Read Write Web (June 12, 2009). Available online: http://bit.ly/d2BHQR (www.readwriteweb.com/archives /social_media_roi_dells_3m_on_twitter_and_four_bett.php).

When measuring the ROI of using social media, we suggest quantifying a goal. Quantify the benefit of reaching the goal. Consider the costs, both in employee time and money (because time is money), and ascertain what it will take to hit that goal. If the costs of reaching that goal are worth it, then jump in and do what you need to do. If the time and money spent does not seem to be worth it upon initial analysis, consider this: your competitors are doing it. Quantify your goals in another way. Will you lose business if you do not make the changes necessary to build a thriving online community? If your business looks to be in danger, make sure you calculate this into your figures. Calculate the negative PR that could result if you avoid changes. What's it going to cost you if you don't? If the ROI of the goal isn't clear, then make sure you include the possible costs of doing nothing.

Community efforts should be a welcome addition to your PR strategy, not a viral addition to your advertising efforts. If your primary aim is merely to drive traffic to your current site, chances are that your community will fail miserably. Creating a MySpace page so that people can friend your new shampoo is not community-building. Lumping this advertising cost into the cost assessment of your community plan is budget mismanagement. Advertising is vital, but don't lump advertising into community budgets, because it's not the same. Viral marketing may very well be important to your company, but this is a part of advertising, and shouldn't be confused with community building. Community building is a long-term effort. If you are having trouble deciding how to allocate your resources, consider the length of the campaign. Community is for the long run, and should be around as long as you are. Advertising campaigns have a finite beginning and end.

In the new world, a place where people are creating connections online and looking to *belong* to a community, it isn't enough to treat your customers like customers, unless your only aim is to sell them something. This is a valuable goal, but a sale is a singular event. In the new world, what you're looking for is 'fans.' Fans come back to you over and over, and they thank you for taking their money. Your customers and your community may be part of your target market, or niche market, or any other kind of market, but it's important to remember that community members aren't only your customers. They're your fans; your community.

Focusing Your Efforts on the Community

Building a community means focusing your efforts on delivering a place that provides value to your community members. Today's web is about users being in control of their Internet experiences. Ten years ago, developing a web presence was more like hanging up a shingle, and even the best efforts were really no more than an online business card. Now there are so many shingles and cards out there that successful companies have to differentiate their businesses. In order to be noticed, your site must provide something that is really useful and worthy of a user's time. Break out of the 1990s mentality and be prepared for the next iteration of the web, which we're already seeing comes with real personalization, real customization, and real value-add to a user's day.

This means you need to take the time to figure out what your community members want, and then you need to give it to them. It's no different than the days of old when we held focus groups, iterated on a product to meet the demands of the group, and held another focus group to see if our product would pass muster. The major difference now, is that learning what today's users want is so much easier, and cheaper. All you have to do is listen.

One of the more interesting sectors in this area is retail fashion. Some of these websites offer rich media that includes different views of products, ratings and reviews, options for sharing via Twitter or Facebook, and even the ability to try on outfits for an idea of what they might look like on someone who has your weight, height, hair color and skin tone. Lands' End was a pioneer in this effort, a decade ago, when they became the first company to use the MyVirtualModel program. According to a 2001 press release from MyVirtualModel, the makers of the technology behind the project, the Lands' End conversion rate increased by 26% and Average Order Value (AOV) increased by 13% during the period between November 2000 and April 2001![6] Those numbers are significant, and that's not even taking into account the probable drop in returns!

---

6. "New Data from Lands' End Shows Value of My Virtual Model™ Technology." MyVirtualModel (press release), Montreal, CA: September 25, 2001.

Why was it so successful for Lands' End? Well, what's the worst part about shopping online? You don't know if what you buy will fit, and even if it does, you can't really tell how it will look once you've put it on. Will this bathing suit make my arms look fat? Will this skirt make my legs look short? These are real factors that impact the sales decision, and Lands' End knew that if they gave their customers some help in that department, it would benefit them as well. Lands' End anticipated customer needs and, in the process, removed a major barrier to sales. Know what customers want and give it to them!

The ability to create wish lists and share them with others is an easy way to do something similar, and it's a feature we see on the most successful retail websites. Some stores take the experience further by allowing customers to rate items, bookmark them, flag the unscrupulous, read the product blog, and share reviews of products they've purchased. Amazon.com has derived much of its success from fostering these community activities.

Case Study: CafePress.com

The CafePress.com website is a great example of a community-minded retailer. The idea behind CafePress is that you can go in and buy products, create products, and sell your own creations, as well. The site features the ability to look at products from different views, and in different colors.

CafePress incorporates several community features into its website:

- Forums

- Chats

- Workshops

- Blog

- Evaluation Panel

- Learning Center

- Planned Events (some of them offline as well as online)

CafePress users are not only members of the CafePress community, but also members of distinct communities within the ecosystem. Each designer has the opportunity to build his own fans, and CafePress supports designers' community-building efforts. While many sites have at least some level of online community building tools, CafePress goes the extra mile to build offline community. Sellers can hold events offline, and surface those events via CafePress. So a seller may hold a local meet-and-greet, and CafePress may hold a larger, site-wide meet-and-greet at a later date, giving fans more opportunities to interact with sellers and like-minded fans.

Their attention to offline community building creates a sense of solidarity and a high level of branding opportunity, as customers openly label themselves as *CafePress customers*. Further, these offline events allow community members to connect on an even deeper level—ultimately resulting in greater loyalty to the community, and the brand. Once you've been to a few CafePress events, you begin to understand what all the hoopla is about. This is real fan-building. Where do you think attendees are going to go to buy their next fun T-shirt, or, even better, where will they look when they're ready to design a shirt for their small business?

Five Things You Can Do to Build Community

It's vital that you understand that your community members will initially give their loyalty to the *community*, and not to your business, your brand, or even to your website. Community members are connected to the community. If you piss them off, they may pack up and leave—and take their friends with them. We cannot overstate the importance of understanding this concept if you plan to build an effective community.

Customers may decide to leave (and they'll do it with very few qualms). Community members, however, are emotionally invested in their communities. They take part in discussions and befriend other community members. They have put time and effort into belonging to the community and they may think twice before packing up their toys and going home.

Customers may help each other by rating a few items, and writing a review or two. Community members, though, take involvement to the next level. They have discussions and debates; they strive for the most ratings, the best reputation, and the accolades you provide for involvement. They share stories and forge lasting ties; they share favorite items with each other and commiserate on related discussions. As a business, you must understand this and promote your website, or even your Facebook page, as a place for community members to build deeper connections to each other. And also, you must reward genuine effort.

In the following chapters, we will take a look at five things that you can do to foster community on your website:

1. Use straight talk.
    Tell it like it is.

2. Use your community members for positive change.

   Invent ways for your community to be more involved and use their information to improve your product and website.

3. Visibility.

   Get out there. Be seen.

4. Tweak.

   Be willing to change things to better suit the needs of your community.

5. Remember the Golden Rule.

   Think about how you want to be treated—then treat your community members the same way.

2 Straight Talk and Using Community Members for Positive Change

In this chapter we will look at the first two of the five things you can do to build a community online: Give them **Straight Talk** and **Use Them for Positive Change**. Both of these offer value for your community and your company.

Straight Talk

When you use straight talk, you tell it like it is. When you are building a community, how you speak to your members demonstrates the kind of community you aim to foster. Your messaging sets the tone, so make sure it's the tone you want to present. When you are straightforward, it communicates to your community members that this online meeting place is one in which sincerity is valued. It helps you convey that you value genuine and real discourse.

Instead of couching your language in terms that may be overly stiff and professional (think PR-speak), go the informal, personal route. You don't sit around with a group of friends and speak as if you are reading from a brochure. Don't use that type of language when interacting with people online. The PR voice should be banished from community areas. If you want to build a real online community, it has to *feel* like an offline community—people sharing thoughts and ideas, or even just sitting around, chewing the fat.

Case Study: Flickr

What *not* to do

Here's the deal: In most circumstances, we like to give second chances, so we'll send you a warning if you step across any of the lines listed below. Subsequent violations can result in account termination without warning.

- **Don't upload anything that isn't yours.**
  This includes other people's photos, video and/or stuff you've collected from around the Internet. Accounts that consist primarily of such collections may be terminated at any time.

- **Don't forget the children.**
  Take the opportunity to filter your content responsibly. If you would hesitate to show your photos or videos to a child, your mum, or Uncle Bob, that means it needs to be filtered. So, ask yourself that question as you upload your content and moderate accordingly. If you don't, it's likely that one of two things will happen. Your account will be reviewed then either moderated or terminated by Flickr staff.

- **Don't show nudity in your buddy icon.**
  Only content considered "safe" is appropriate for your buddy icon. If we find that you've uploaded a buddy icon that contains "moderate" or "restricted" content, we'll remove the buddy icon, moderate your account as "restricted" and send you a warning. If we find you doing it again, we'll terminate your account.

- **Don't upload content that is illegal or prohibited.**
  If we find you doing that, your account will be deleted and we'll take appropriate action, which may include reporting you to the authorities.

- **Don't vent your frustrations, rant, or bore the brains out of other members.**
  Flickr is not a venue for you to harass, abuse, impersonate, or intimidate others. If we receive a valid complaint about your conduct, we'll send you a warning or terminate your account.

- **Don't be creepy.**
  You know the guy. Don't be that guy.

- **Don't use your account to host web graphics like logos and banners.**
  Your account will be terminated if we find you using it to host graphic elements of web page designs, icons, smilies, buddy icons, forum avatars, badges, and other non-photographic elements on external web sites.

- **Don't use Flickr for commercial purposes.**
  Flickr is for personal use only. If we find you selling products, services, or yourself through your photostream, we will terminate your account. Any other commercial use of Flickr, Flickr technologies (including APIs, FlickrMail, etc), or Flickr accounts must be approved by Flickr. For more information on leveraging Flickr APIs, please see our Services page. If you have other open questions about commercial usage of Flickr, please feel free to contact us.

Flickr can give us many great examples of straight talk. My favorite, the 'Flickr Community Guidelines' are easy to read, and use plain language. Typical Terms of Service agreements appear to be written in the hope that people will not read them. Flickr's Community Guidelines,[7] on the other hand, beg to be read.

Some examples of Flickr's straight talk:

"Don't upload content that is illegal or prohibited."

and

"Don't vent your frustrations, rant, or bore the brains out of other members."

Under each title, Flickr includes a straightforward (and sassy) explanation. For the venting guidelines, Flickr offers this:

"Flickr is not a venue for you to harass, abuse, impersonate, or intimidate others. If we receive a valid complaint about your conduct, we'll send you a warning or terminate your account."

The language is clear and concise. It sets forth expectations, and it describes punitive actions, leaving no doubt what Flickr stands for, and what the site administrators will do if you break the rules. The Flickr Community Guidelines also display the site's hipster attitude. Flickr tells it like it is, and does it with flair. The best guideline from Flickr:

"Don't be creepy. You know the guy. Don't be that guy."

I do know that guy, and I don't want to be him!

---

7. Flickr Community Guidelines, http://www.flickr.com/guidelines.gne.

In addition to giving community members and visitors a clear idea of what to expect from your website, straight talk also endows you with a reputation for honesty. When you tell it like it is, you set yourself up as a trusted source of information. In order to successfully build an online community, it is important for people to believe in you and your message. This means that sometimes you have to take the blame for mistakes, owning up and explaining how you are going to make it right. If you do this promptly and with as much transparency as possible, your honesty will endear you further to your community members.

A Few Thoughts about Dissembling

The opposite of straight talk is dissembling. Online community members are wise enough to know when you are being disingenuous. If you are trying to talk around an issue, they will know it, and they will resent your obvious dishonesty. No one likes to feel as though someone they trust (or want to trust) is giving them the runaround.

Dissembling isn't going to help you build a better online community. Instead, it will settle you firmly, in the minds of potential community members, as someone who shills for their own benefit. With millions of people online all the time, and thousands looking at your website, someone is bound to notice if you are not what you seem. When that someone does find out that you have been dissembling, he or she will broadcast your perfidy as far and as wide as possible.

Attract Fans to Your Site

When your community members trust and like something about your website, they will be more willing to share it with others—and encourage additional participation. This fact presents a great opportunity for you to grow your online presence and community. In many cases, you can use your fans to help publicize your company.

You don't want visitors to your online community. You want active participants. You want fans. Fans are community members who are involved and engaged. They love you, they love your product, and they love being a part of your community. You do not need to encourage a fan to speak positively about you. They do so because they are interested in what you have to offer, and they like it so much that they want to share it with others. You need to develop a community that creates fans, rather than a community of visitors who drop by every once in a while. You want community members who bring *their* friends to join the community as well.

Here are some standards that you should consider when trying to attract fans:

- Create useful content.

- Offer a quality product or service.

- Show you care about community members.

- Include rich media and anticipate the community's wants.

- Provide a place for discussion.

- Update regularly.

- Consider swag, tchotchkes and blatant praise for über users.

- Make it easy to subscribe to the content on your site.

- Make it easy to share content off-network.

- Invite community participation often.

- Compile shared content into an easy-to-use knowledge base.

- Present a coherent theme.

The idea is to make your online community somewhere fun and useful to be online. If you can drum up some fanboys,[8] all the better.

---

8. "Fanboy/Fanboi," Wikipedia, http://en.wikipedia.org/wiki/Fanboy.

Case Study: Apple

One of the greatest examples of cultivating fandom is Apple. Apple fanboys are spread out all over the world. They are so enthusiastic about Apple products—from the MacBook redesign to the latest method for hacking the iPhone—that there are thousands of blogs and discussion boards devoted to Apple and its products.

Apple has an official community where users can meet to help each other and get their questions answered. But the real success isn't their apple.com community, but fans who share their knowledge off-network. The moment Apple announces almost anything, the buzz becomes almost overwhelming because fans are enthusiastically arguing about it. You can't even read an article online that mentions Apple, or its arch nemesis, Microsoft, without the comments deteriorating to a Mac vs. Windows debate. Apple has some very motivated fans.

Even Apple haters help Apple. Apple haters are a community them-selves, and their responses help keep the company in the spotlight. Apple fans, and their anti-fans, drive word of mouth and online buzz. Remember when the iPhone was initially released and the stories of jailbreaking[9] began to get widespread attention? It was on all the major television news shows, magazines and newspapers, and even crept into water cooler talk. This buzz grew from only a few initial fans, sharing how they'd gotten their iPhone onto other networks, cir-cumventing the AT&T-only protocols. Everyone was talking iPhones. How many additional iPhones were sold because of the word-of-mouth buzz created by this story? Thanks to die-hard fans and Apple-focused communities, more people were introduced to the iPhone.

But how did they build this great fan base? I hope that we all under-stand that Apple's lovefest did not happen by accident. First, Apple has amazing products, and it helps your case when you build things

---

9. Associated Press, "New Jersey teen cracks iPhone network lock," MSNBC.com, http://www.msnbc.msn.com/id/20424880/.

so well, that work so intuitively, that people decide they simply can't live without them. Build the best mousetrap. Don't expect to build fans with a product or site that has no real value.

But you need more than just a great product, as Sony can attest. To build fans, you need a dedicated, long-term strategy and support from the highest level of the company. Apple has always made their best effort to understand their customers, and they've had a strategy for building fans for decades. Apple has had a team of evangelists since the 1990s, most notably the former Apple evangelist, Guy Kawasaki. Evangelists cultivate that intense love, and because Apple continues to build solid, innovative products and provides genius-level support to back them up, the company has more cheerleaders than it can count. Apple is one of the few companies who have understood online community from the get go; this understanding has paid the company back in spades.

Using Your Community Members

But Apple isn't perfect, of course. Like all companies, Apple is comprised of fallible human beings. Sometimes even *they* muck it up. Luckily, with a fan base like theirs, even missteps can turn into blessings in disguise.

Let's return to the iPhone jailbreaking example. From the beginning, the major complaint from iPhone owners and prospective owners was that the iPhone could not be used on a customer's carrier of choice. And, while Apple had always warned that hacking the iPhone so it could be used with other carriers might damage the phone, the company didn't expressly prohibit the practice at the outset, likely hoping to retain their most fervent fans, who would hack it and bring it to another carrier. These people were not part of the majority, they were fringe users, and Apple did not want to offend them, nor shed additional light on the unfavorable truth of the exclusive carrier agreement, so they all but ignored the jailbreakers. Perhaps they even secretly applauded the efforts of fans who'd, in essence, given the iPhone additional coverage in the media.[10]

The silence made most people complacent, so they were shocked when Apple cracked down by having its firmware updates disable iPhones that had been hacked,[11] essentially turning the device into a sleek and expensive brick. Pressure from AT&T probably accounted for the abrupt change in policy, and was a powerful deterrent to would-be jailbreakers. For many companies, this would be, at the least, a major PR nightmare. For Apple, it was a minor story. Without the Apple community's already solid love and support, though, this could have gone much differently.

Consider for a moment how Sony is viewed—as a company that takes proprietary accessories to an absurd level. Many hate Sony for that reason, and will buy a competitor's product to avoid having to use only genuine Sony cords, memory cards, and other accessories. Apple is at least as guilty of this. With Apple, it's not only that finding non-Apple products that work well with your devices is difficult, but even their own accessories aren't interchangeable. Remember, their own iPod car chargers and cords don't even charge their own iPhones! Any upgrade almost always requires additional cords, ear buds and chargers. They take proprietary to a whole new level. Yet, oddly enough, Apple is only vilified in hushed tones, because we know we can't live without their products and because so many of us are diehard fans. Sit back and consider that the iPhone is so perfect, that not only are we willing to put up with having to buy proprietary accessories, but we barely complain about the horrible service from AT&T.

Fans make all the difference. You can enjoy similar success with your community members as well, if you treat them properly and your product or service deserves fandom. Encourage community members in their efforts. Provide places for them to meet, and give them reasons to share your message with others. Tools that you can use to encourage your fans are usually discussion based.

- Wikis

- Message Boards

---

10. Michael Krigsman, "iPhone unluck hack: cooked Apples or not?" ZDNet.com, http://blogs.zdnet.com/projectfailures/?p=390.
11. Ryan Fass, "Latest iPhone update and Apple's response could spell trouble," ComputerWorld, http://blogs.computerworld.com/node/6287.

- Blogs

- Email Lists

- Social Networks

- Podcasts

- Suggestion Areas

Think of ways to stay connected to your community members when they leave your site. Do you have a Twitter account that users can subscribe to for the latest news? Can your community members participate in forums? If they can, is it a one-way channel, with your fans talking to you only, or do you actually respond? Do you offer how-to video tutorials? Have you ever asked highly active members of your community to make some for you? Do you put together funny or useful (or both) podcasts? To bastardize Kennedy: Before you ask what your fans can do for you, ask yourself what you are doing for your fans.

Of course, some of these suggestions won't fit every business. It is important to adapt community-building tools to your business. A podcast about carpets, in general, might fail; it's boring and doesn't have wide appeal. However, a carpet company might create or sponsor a lifestyles podcast, thereby creating a beneficial offering for its community. If it's useful and/or clever, your fans will share it with people they know, spreading your brand name and fostering brand recognition. That's the end goal, isn't it?

Addressing Potential Problems

You can also use your fans as watchdogs. Your community members are your knowledge base and your eyes and ears. They help each other, and they warn of problems and potential problems. You should pay close attention to what your customers are saying online. This can alert you to specific problems and issues that need to be addressed.

Community members can also provide information and feedback on an even more practical level. You can take your community members' insights and make your product or service better. Their feedback can also help you improve your website and grow your online community.

The key, though, is to nurture a constructive environment in which community members are invited to share, without worrying about being penalized for valid criticism and suggestions.

Of course, you can always ignore criticisms from community members. Let's see what that does for you. A very bad example on how to manage online feedback is offered by Coca Cola. Sit down at your computer and do a Google search for 'Coke.' You'll see that 'Killer Coke' remains on page one of the search results. If you don't know what that means, take a minute to read what it's all about.[12]

Now, would you want something so damaging on page one of the search results for your BRAND NAME? And this isn't a new thing; it's been like that for at least five years! Don't ignore negativity, and don't fight it with a similar response. React to criticism as an adult would—learn from it and act on it. And don't make the same mistake again.

Coca-Cola | Facebook
Bienvenido(a) a la página oficial de **Coca-Cola** en Facebook. Recibe contenido exclusivo e interactúa con **Coca-Cola** sin salir de Facebook.
es-la.facebook.com/coca-cola - Cached - Similar -

Killer Coke ☺ ★ ★ ★ ☺ ▲ gavinhudson + 2 ☺ Activism
There are undisputed reports that **Coca-Cola** bottling plant managers in Colombia, South America, allowed and encouraged paramilitary death squads to murder, ...
www.killercoke.org/ - Cached - Similar -

Why Does Coke From a Glass Bottle Taste Different? | Popular Science ☺ ★ ★ ★ ☺ Science
"The great taste of **Coca-Cola** is the same regardless of the package it comes in," ... The people at **Coca Cola** are full of BS. Coke in a glass bottle tastes ...
www.popsci.com/.../why-does-coke-glass-bottle-plastic-bottle-and-aluminum-can-taste-different - Similar -

KO: Summary for COCA COLA CO THE - Yahoo! Finance
Aug 4, 2009 ... Get detailed information on **COCA COLA** CO THE (KO) including quote performance, Real-Time ECN, technical chart analysis, key stats, ...
finance.yahoo.com/q?s=ko - 31 minutes ago - Cached - Similar -

Coca-Cola Television Advertisements: Home Page
Coca-Cola television advertisements from the motion picture archive at the Library of Congress.
memory.loc.gov/ammem/ccmphtml/ - Cached - Similar -

---

12. http://www.killercoke.org

3 | Visibility

The previous chapter addressed the ideas of speaking like a human being to your community and using your community members for good. While these can solidify your community members and create loyalty, you also need to be among them, mingling and socializing. In order to grow your online community, you need to increase your visibility. You need to let other people know you are there.

Community experts use several ways to increase a company's visibility and draw attention to their online community, and you should pay close attention to what they do. In this chapter, we will look at blogs, social media networks, conferences, events, message boards and even the mainstream media. All of these provide outlets for you to increase your visibility.

Blogs

A blog can be a very useful tool in terms of raising your visibility. It allows you to easily update your website regularly, while providing valuable content for your community members. Blogs also do very well in search results, because of their heavy reliance on text (search engine candy). Not only do many blogs show up in main search results, but Google also has a way to limit search results to blogs only.

If you want your blog to be more visible on the Internet—drawing more people to your online community—you need to keep a few things in mind:

1. Post regularly, at least in the beginning. Regular posts (3–5 a week) signal an active blog and community. Your site will be more attractive to prospective readers if you regularly provide new content.

2. Speak in a human voice. Sure, we covered that in the last chapter, but it's important to emphasize that a blog is definitely a 'common man' type of outlet. It's a casual writing tool, and people reading your blog don't expect content that is sales-heavy, nor do they expect content that is too academic (there are some exceptions to this rule, of course, depending on your customer base). For the most part, keep it casual and add some humor, when possible.

3. Keep it valuable. A blog should not be all about your company. It should be more than one sales pitch after another. Include useful information and valuable content. If you offer useful and/or entertaining content, people will keep coming back. They'll want to be a part of your online community.

4. Allow comments. You'll have to have some sort of moderation (no one likes spam or inappropriate comments), but you shouldn't just delete the negative stuff about your company. Members of an online community like to have their say. Paying attention to what they write can give you great insights as to what motivates your readers. Negative comments also give validity to the positive ones, so don't remove anything unless it's offensive, private, or spam.

5. Use tags, keywords and good headlines. Keywords are important, when using a blog, to boost your online visibility. But you also need to make sure that your tags and headlines include relevant keywords. While a catchy and clever headline may be 'fun,' if it doesn't have keywords that are relevant to your subject matter, it's not as valuable.

6. Dress it up a little. Add images, audio and video to your blog (with proper permissions and attributions when necessary). Additionally, you might vary the flavor with posts of different lengths. This will make your blog visually appealing and create an atmosphere of realism. If every blog post is between the marketing department's specified 500–750 words, are you saying something important or just shilling?

Social Media Networks

Social networks have been gaining in popularity with even the most mainstream of folks recently. Facebook, LinkedIn, and Twitter are just a few examples of sites that now draw mainstream users. Setting up an account on some of these services can be a great way to increase your online visibility. Sites like these allow you to meet people and share information. It can be even more effective if there is a social network that caters to your particular niche. Real estate companies, for example, have the social media network ActiveRain. Facebook is a site built of many smaller communities. You can choose to build your own fan page or take part in a Facebook group that is made up of people you want to reach.

Case Study: LinkedIn

For businesses and business people, LinkedIn is a key social site. It is designed to be a professional network, and people use the site to make business connections and to network. Here you can do better than just find friends and connect with like-minded people—you can search LinkedIn for potential business partners and for employees.

LinkedIn has a 'connections' feature that allows you to find people you know, and then look at people *they* know. If you have a good profile, it is possible to grow your online community when connections are made through people you know. LinkedIn has the potential to help you build your online community in a way that is full of value for the business professional.

LinkedIn has also introduced a way to follow companies. Consumers can follow companies that they would like to learn more about, or even find all job openings in that company. This feature is still new, but already businesses are using it to increase the number of people who hear their message.

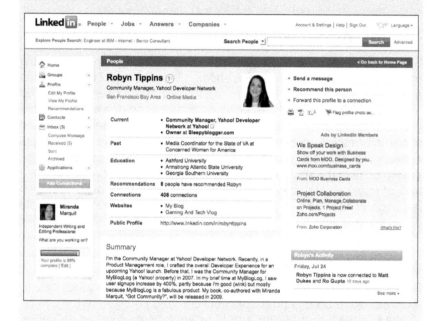

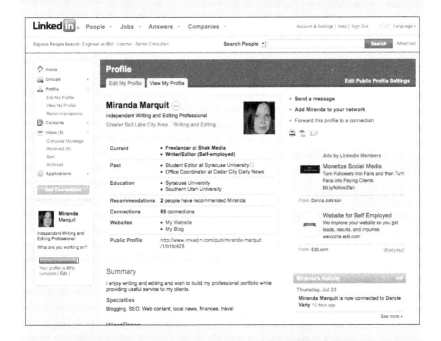

Your profile is an important part of raising your visibility on the Internet. Make sure that your profile is interesting, yet appropriate. Include keywords in your profile, but take care that they make sense (more on this later). Having a picture helps, as does a bio that accurately describes you.

Be aware that you do not have time to join every social network, nor would you want to join everything. The truth is, developing a good profile and keeping up ties in social networks can be time-consuming. Think about which groups you want to target the most, and then join the sites that they are most likely to be using.

Conferences and Events

A fabulous way to enhance your visibility is to attend select, relevant conferences and events. There are several conferences aimed at getting people together. The BlogWorld Expo in Las Vegas is one of the most popular events based around bloggers and those involved in social media. When you attend such a conference, you can meet new people and get the word out about your blog and your company. When you meet people—especially if they already have followings—you stand a good chance of getting in front of their readers, thereby gaining access to another audience. Another good example of a blogging conference that is great for visibility is the BlogHer conference, which focuses on blogs by women, or those focused on women's issues.

South by Southwest (SXSW) also draws people from all over. Even though this is primarily a huge music and film festival, it attracts a diverse audience. Indeed, the social network Twitter initially took off because many people at SXSW were sending in updates from the event. (Note: Just because these conferences represent those favored by the authors, it doesn't mean they will be useful for you and your company. Do your own research and make sure that the cost of the event is worthwhile. Don't waste your time and money going to events because they are popular; only spend the time if it's really worth it for you.)

Beyond conferences, you might also hear, through your connections on social sites, word of meetups. If there are meetups in your area, give them a try. There's very little risk because they're usually free or very cheap, and the worst thing you can find out is that you've wasted an hour of your time. The investment in these types of groups is so small that getting a significant ROI is almost guaranteed. Making face-to-face connections is a good way to foster online relationships. When you meet someone offline, s/he is more likely to come find you online. And, if you have a good website and online community, you are likely to gain more members as a result. To find relevant meetups, try searching by location and topic on Upcoming.org and Meetup.org.

Message Boards and Forums

Message boards and forums can be great ways to raise visibility. Additionally, they can foster conversation in your online community, thereby increasing engagement. You can use these tools to get people talking about issues related to your company, or use them as sounding boards for your ideas. The best part is that they provide a place where community members can interact with each other—all on *your* company website. Just remember that for message boards and forums to work, you need to be an active participant as well.

You can use message boards that are on your own site, as well, to converse with your audience. Robyn is Community Manager at the Yahoo! Developer Network (YDN). YDN uses both onsite and offsite message boards. Their onsite message boards are the hub of the YDN communities. They allow YDN to help developers debug their code, get started using Yahoo! APIs, and give the team valuable feedback on products.

Mainstream Media

It may seem strange to include the mainstream media as a key to building your online community. But it can help you reach out to others through non-Internet channels. If you put ads in your local newspaper, make sure you include your web address. The same is true of television and radio ads. If you buy a yellow pages listing, make sure that your online community location is included with that as well, even if your community is located on Facebook or some other social site, and not on your own site. Many people like to check businesses out online *before* they visit them, even if they are local establishments.

Case Study: The Local Restaurant and the Blog

Miranda says: There is a relatively new restaurant in my town that doesn't have a website, but does advertise in the newspaper and on the local TV station. When my husband and I went to try it out, we were disappointed in the food. I wrote a review about it on my personal blog. Many visitors to my blog say that they saw an ad for this restaurant in the local paper, and then looked it up online. In a search, the first thing that comes up is my negative review. It probably isn't destroying business, but this restaurant, if it had a website, could include it in their ads, sending people directly to their own site, and circumventing my review. This might net them a few more customers.

It is true, though, that mainstream media advertisements can be pricey. Television and magazines can be especially expensive. But if you start local, you might be able to keep the price down. Plus, many mainstream media outlets have their own websites and offer online advertising at a discounted rate. You might be able to advertise on the website of a popular local TV news or radio station to raise your visibility and attract new online community members.

Yelp is now the go-to place to check out local businesses. If you are looking for a great Chinese restaurant or take-out, you can go to Yelp, do a quick search, and see ratings on the prices, food, and hygiene and even lengthy reviews as to the service and quality of the food. Reading through the reviews can also give you ideas as to what you should order, and if certain times of day are busier than others. And, on Yelp, everyone from plastic surgeons to nail salons are locally reviewed. There have even been lawsuits as to the damage that Yelp reviews have done to a business' livelihood. Have you looked at your own company's Yelp review?

Now that you are more visible, go see what there is to see.

Part of visibility is 'visiting' others online. A 'look at me, look at me' mentality won't do you any favors. You can't just set up a profile on a social site or start a blog and expect people to just show up. A few will, but you need to go out and see others as well.

After you have a few posts on your blog, go read others' blogs and leave insightful comments. Find message boards and forums in your company's niche and sign up *and participate*. Set up a profile on a social network and then find the profiles of others with similar interests. When you meet others at conferences or events, make sure you pay them an online visit. If you want to build your own online community, you need to be a part of the online community yourself.

Many forums allow a signature. Create one that includes a link to your website or blog. When you leave a comment, most forms ask for your web address. Make sure you include it. This is very important. You want people who find your insights interesting and relevant to be able to easily find (and join) your online community. Of course, this assumes your insights are interesting and relevant. If they aren't, start now to improve the way you present yourself in comments and on forums.

Warning: Spam

When you are out there 'visiting' others in cyberspace, you need to make sure you are following the basic rules of online etiquette. It should go without saying that you should avoid profanity and personal attacks on others, as well as try and maintain a professional image for your company in the online realm. But what isn't so obvious is how important it is to avoid spamming on your comments and forum posts.

Spam basically amounts to blatant advertising for your company. It is when you go and visit websites and message boards for the express purpose of leaving your web address. Spam does not contribute something substantial to the conversation. This is why you should restrict your website links to the form and to your signature. The content of your contributions should be thoughtful, relevant and useful. If you consistently show up to do nothing beyond promote your website, it will be noticed, and you may be shunned (as you should be).

Spam is *not* the way to build a thriving, growing online community, and it's a royal pain for the rest of us who really want to take part in these communities. If you can't say something of value, don't say anything at all.

4 Tweak Your Online Presence

As we've pointed out, the Internet has evolved beyond static HTML, and you need to evolve as well. If you want to build a successful online community, you need to keep changing and updating your website. If you see something wrong—or something that can be improved—you need to tweak it.

Tweaks aren't major overhauls; they are small changes made to add value to your current offerings. They don't take months to implement and they don't need a staff of 50 engineers to accomplish them. Some examples of tweaks include:

• Rewording help pointers in response to customer questions

• Designing all pointer icons similarly

• Changing all URLs to 'pretty URLs'

• Simplifying a confusing login process in response to customer feedback

• Adding Title HTML tags to all images for better SEO and for the aid of those members with slow connections

- Adding a printer-friendly link to pages that are frequently printed by your audience

When you tweak items on your website, it's not about change for change's sake. In fact, it should rarely be about change for change's sake. The most effective changes to websites are those that are focused, and have a purpose. Before making changes to your website, you need to ask yourself why you are making these changes. Is it because you see a need and want to fill it? Is it because your change will increase the SEO of your page, or make it speedier? Are you changing your site because all the other hip/cool companies are adding glass and reflections to their page and you want to be hip/cool like them? None of these are wrong in themselves, but all would be if they hampered the ability of your users to use your site. The best changes can be bad if they are implemented poorly. Usability is paramount. All tweaks should happen in a way that allows your website to better serve your audience, customers and community.

When you tweak your website, there are four main positive reasons to change:

1. Changing in response to feedback makes your community members feel good about being on your site. They feel as though you listen to them and care about them.
2. Tweaks can make it easier for users to navigate your site, allowing for wider access to your products and services.
3. Customer service inquiries can be reduced (especially if your FAQs are properly tweaked), freeing you up to make more money—and freeing up your bandwidth for more productive transactions with community members.
4. Some changes can increase the amount of time community members spend on your website. More time on your site often leads to an increase in products and services used/sold.

Tweaking the Little Things

Tweaking is all about making small, effective changes to your website. These little things, though, can add up to a big difference. Some of the things you can regularly tweak include frequently asked questions (FAQs), tutorials, menus and content.

FAQs: Tweaking your FAQs can be a great idea. Do you receive a large amount of inquiries on a specific subject or item? If so, check your FAQs. Do you address it clearly in this question-and-answer section? If you don't, then modify your FAQs so that they really do address common concerns and issues. You can add more questions and improve explanations. If there are questions that are almost never asked, you can remove those from the FAQs page, making it less cluttered while leaving room for questions that more people are concerned about.

When creating your FAQs, remember some of the items we discussed in the previous chapters. Straight talk is especially important on the question-and-answer page.

Tutorials: So many businesses put up tutorials (or even FAQs) and then forget about them. But the fact of the matter is that you can't just do something once and leave it there. When new information comes out, you should change your tutorials to reflect that. Whether you have written tutorials or video tutorials, make sure you periodically update them.

In fact, if all you have are written tutorials, alter them by adding images and videos. Visuals are more attractive to community members, and videos can make instructions even easier to understand. Adding images or video can be applied to more than just tutorials though. Consider adding these visual elements to product descriptions (of popular products and services), as well as reviews. Amazon.com offers community members the chance to upload customer images and even video reviews of products. This is a great example of tweaking the website and increasing community involvement.

Menus: This is a biggie. Some business websites are a real pain to navigate. If your community members can't find what they are looking for, they are more likely to look elsewhere for what they need. Whether it's finding 'contact us' information or locating a specific product, you need to make sure your website navigation menus are easy to use—and options are easy to find and understand.

One great addition to the menus is a visible 'search' function. If you make your entire site searchable (including FAQs and videos), your community members will be able to easily locate what they are looking for. Tweaking your website so that it has a good search function is a great way to keep your community members happy—and that makes them more likely to return to your website in the future.

Content: It is important to continually update your content. Change articles and blog posts when the information changes. Add new content regularly. If your forums are slow, think of something interesting to add and start a new thread yourself.

Also, tweak your business website so that it has different types of content. Add podcasts of interesting information and how-to's. You can integrate live webinars that allow you to interact with community members in real time. Add social tagging functions to your website so that community members can share your content. Small tweaks to your content can keep your web portal fresh and interesting.

Caveat: All of these can be overwhelming all at once. If a customer is used to a site with no community features, it's rarely advisable to add video, reviews, ratings, podcasts, etc. all at the same time. In fact, some sites, depending on audience and content, will not benefit at all from these changes. You have to decide what is best for your company. If you need help, do research and/or hire an expert to direct you as you strive to do what's best for your company website.

Tweaking the Big Things: Website Overhaul

Okay, so a website overhaul isn't exactly tweaking. It's more like reinventing. But, in some cases, it might be really necessary. If you need to adjust the look of your website to bring it into this decade, don't skimp. Figure out what you need in terms of resources in order to ensure that your website loads quickly and meets the needs of your community members. Consider hiring a user interface (UI) designer to help you determine where the content is best placed. They understand user behavior better than the rest of us and can help you increase use of key areas, and decrease registration drop-offs.

Before undertaking a website overhaul, make sure you study the issues. You need to figure out which things to bring over from your archives. Searchable content from the past (evergreen content) is something that should be present on most websites, and if you point to older content with keyword-heavy title links, you can give certain areas on your site a great SEO boost. You should also consider different designs and schemes for your new website. Navigation should be a consideration, and you'll probably go through dozens of wire frames before you decide on a structure and design you know your audience will appreciate.

Along the way, you need to make sure that your community members know what's going on. This is vital if you decide to change your domain name (which you should only do if the alternative is prison or something else similarly horrid). Make sure you invite community members to the re-launch, and encourage them to spread the word. Ask for their feedback early on, but remember to take it with a grain of salt. They may have great ideas, but often the most vocal members (i.e. those who respond to surveys and feedback requests) may not be windows into the souls of the majority of your users. Just because a few people tell you that you need to change something doesn't mean that most of your users want, or would even appreciate, that change. Also, regardless of customer feedback that a given change would be good, if that change impacts your business in a bad way, it's still not a good idea. When we say listen to them, we do not mean use them as your business advisors. Gather all feedback, research, consulting advice, etc., and then make your own decisions. Everybody's got an angle, and you are the only one whose angle is to help your business succeed.

How Do You Know What to Tweak?

Many businesses aren't sure what they need to tweak on their websites. Luckily, you have a resource readily available to you: your community members! When you have an online community, members usually like to feel as though they are heard, and that they make a difference.

When eMoms@Home decided to rebrand, Wendy Piersall, its Founder, reached out to her community members, asking them to vote on different new names. The result was Sparkplugging.com. Because she kept her community members in the loop, including them in the process, most transferred their loyalty to the new brand.

Community members can provide great ideas for new content, offer constructive and helpful ideas for improving products and services, and even let you know when they would like something specific done. When you make your website one of safety and openness, community members feel comfortable sharing their insights. They know they will be taken seriously, and you can improve your business website.

Other places that can offer information on what to tweak with your website include sites and blogs in your niche, news stories, the latest developments in the industry and market information on what your community members want. The bottom line is that you need to keep making changes to your website. No one is perfect, and neither is any website. You need to be constantly improving and tweaking your site for the benefit of your community members. Remember: for many businesses, after making money, the most important goal is having happy and enthusiastic community members. Do whatever it takes, within reason, to make the necessary changes to please your community members and turn them into fans.

5 The Golden Rule

A vital point to remember when building an online community is the Golden Rule. Most likely, you learned this rule as a child. 'Do unto others as you would have them do unto you.' In America, where much of the population follows the tenets of some form of the Christian faith, the Golden Rule manifests itself in Matthew 7:12. But all religions, and humanist atheists, recognize the validity of the Golden Rule. And the Golden Rule is one of *the* rules for building an effective and organic online community. The Golden Rule of online community itself consists of five basic components: bribes (yes, but don't freak out), speedy response, accountability, user focus and simplicity. Treat others how you would like to be treated, and you should do just fine.

Bribes

Bribe is a nasty word. We know some of you cringe just seeing something so crass even grouped into the Golden Rule example. But we want to be clear: giving a customer or visitor a freebie usually creates a happy feeling that you want them to associate with your company. Admit it; you like to get stuff, right? So do visitors to your website. Your customers want added value—and that's really what you owe them. Just as you look for the best deals and desire 'extras' in your dealings with other businesses, web users want a reason to pick you above the competition.

Bribes in the sense of online community do not have to mean cold, hard cash. In fact, we think it must be said that cash and rewards can actually backfire. The credibility of your advocates can be called into question if it gets around that you are bribing them with cash (paid comments or blog posts are great examples of what to avoid if you value your credibility). Giving someone money or gifts in exchange for positive feedback not only calls into question their praise, but it means that in the future, they will expect it; their positive feedback will always cost you something. Along a similar psychological vein, we also want to point out that if someone gets paid when they praise your product, their own positive outlook on your company will diminish. They'll begin to doubt that your product is really all they touted it to be if you felt the need to offer them money because they said so. Successful companies know how to give something away and come away looking good, not tarnished.

Non-monetary freebies, such as swag (branded items that include t-shirts, stress balls and key chains) and invite-only events add value to your community without compromising your advocates. Teasers for future subjects and tutorials can also build interest and increase return traffic, building repetitive journeys back to your community. Status upgrades within the community, free premium accounts, or the chance to guest post on the community blog are also great bribes for stellar community members. We hope it goes without saying that these bribes should be given only to your top tier community members, and not to every guy or gal who visits and comments. Make the really special people know you think they're special and you'll build loyalty to you and your community. The most successful companies tend to their base and the rest of their users follow.

Case Study: Travelocity

Robyn says: Value-added programs that you offer on your website can be effective in nurturing a community that people want to be a part of. Travelocity offers a superb example with its VIP service. Value comes via the VIP-only phone number that offers help 24/7 for reservations. Of course this encourages users to choose Travelocity when making their reservations, rather than searching other reservation websites. When I first became a Travelocity VIP, I was thrilled, though that is a little embarrassing to admit. The marketer in me knew this was a clever campaign by a company that wanted to boost the rate of return of their most frequent travelers, but I still have to admit that I did *feel* special. They placed a coveted gold crown icon next to my name, and who doesn't want to feel a little more like royalty? The special discounts were nice too, and they keep me coming back to Travelocity for most of my travel. Many of the travel plans I book for my extended family and even a few friends are done because of my VIP status on Travelocity. I've even booked a trip to retain my coveted status, in spite of knowing that this status was purely designed to trick me into doing so. All it took to make me completely loyal to Travelocity was to make me feel like they appreciated my business, more than they appreciated their normal, sporadic travelers' business. They made me feel unique and I returned their goodwill with thousands of dollars in business each year.

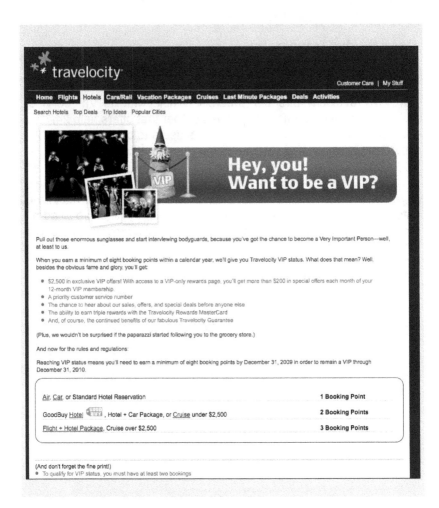

Speedy Response

What's the thing that annoys you most when you call a company? I'll bet it's waiting to talk to a live person. All of us are busy. We prefer to give our business to companies who respond to us efficiently, dealing with whatever problems we have quickly and painlessly. The Internet has only reinforced this desire. Your online community (and eventually your company website as a whole) will become dead in the water if your response time is slow. Your customers deserve a fast response, whether they fill out a contact form or post on a forum. Determine to give them the response they deserve. After all, in the virtual world, a bitter customer can take his or her message very far, very quickly. Don't give community members something negative to talk about, especially when it's so easy to cut your response time. Just remember how quickly you expect to be answered and make sure you respond just as speedily to the people who pay your bills.

Answer emails quickly. Don't make your community members feel like their email is a message in a bottle that will never get attention. There is absolutely no reason an email can't be answered within one business day. If your business isn't scaleable enough to offer a timely response, I suggest you take the 'email us' link off your website. It's just plain cruel to ask them to send you a message, if you can't respect their time and energy enough to answer punctually.

Respond to message boards and blog comments. No, you don't have to respond to all of them, but enough of them that it's apparent they are being read. If you have 200 comments, all concerning a faulty feature on a product, it's unnecessary to repetitively answer each comment. However, make sure you address the issue, and check back often to make sure that the answer is reiterated if the message hasn't sunk in. One caveat: never cut and paste an answer that you have previously given. Your customers aren't morons and they won't appreciate your treating them as if they are. Dignify their concerns or questions with a real response, every time a response is necessary.

Moderate your blog comments. Moderation keeps sensitive customer information out of the public view when a misguided customer uses your blog comments to request assistance, complete with their username and password. Moderation removes comments that are completely off-topic. It keeps spam out of the conversation. A blog that's full of spam and inappropriate comments isn't credible, because it's clear that no one is policing the responses. Moderation also protects you from the liability you will face if a comment on your blog is illegal. While giving your customers a way to respond to your blog posts quickly is vital to the success of your blog, **speediness in this area should not come at the cost of your company's credibility**. That said, moderating comments reasonably quickly is crucial. A popular blog, with dozens of comments per post, requires more than an hourly moderation. A blog that only gets a comment or two per week can be served by daily moderation. Whatever works for you and your customers is adequate. Set up your company's community blog to notify you when comments are posted. Nothing is worse than finding that comments have been stuck in moderation, awaiting your approval, for days. Indeed, unresponsiveness at this level can cost you customers, and deservedly so.

Police your message boards. We won't rehash what's already been communicated, but suffice it to say that a board that gets no attention from its owners tends to get really seedy, really quickly. If you don't want your message boards to be a virtual breeding ground of warez links and porn clips, keep an eye on your boards. Remove inappropriate content quickly and stay involved in the discussions.

Be speedy in your moderation and in your responses. Timely communication tells your customers that they are important to you and that is exactly the message you want to send. If a question or concern comes in via email, or even on the forum or blog, and you do not have the resources or ability to address it immediately, be sure that you respond anyway. Briefly acknowledge that the customer's message has been received, and that representatives are working on it. Keep them abreast of new developments as the problem unfolds. Err on the side of over-communication rather than appearing to ignore the problem. Remember, if you don't have the resources to manage a blog or message board, don't create them. Having no community features at all is better than having poorly managed ones.

Case Study: Digg

One of the most interesting cases of community backlash occurred back in early 2007. Digg.com removed a controversial sixteen-digit key that unlocks HD-DVD titles.[13] The key allowed users of the Linux operating system to view HD-DVDs, something they had previously been unable to do. However, the key could also, theoretically, open the door for intellectual property theft and the powers that be at HD-DVD were not too keen on that key getting out. Digg received a cease-and-desist order, and chose to remove the code rather than leave themselves open to liability. Stories with the code were taken down.

Soon, stories criticizing Digg's decision appeared. In some cases, Digg buried stories about the HD-DVD key—and criticism of the moves taken by Digg quickly occurred. But it didn't stop there. After Digg buried the first stories, more began appearing. They were buried of course, but the stories just kept coming, and the inflammatory comments that accompanied every one of those posts got worse. At 1 p.m., on May 1st, 2007, the CEO responded on the blog and explained the situation, but Digg users didn't give up. They persisted in posting HD-DVD keys by the hundreds, and Digg just couldn't hide them all. At one point every story on the Digg homepage was an HD-DVD story with the key in the headline. There was no stopping this key from getting out. The community had made its point clearly.

Finally, at 8 p.m. (within 24 hours of the first post of the key), the founder, Kevin Rose, posted that they were not going to bury any more stories with the key in it.[14] In fact, he posted the key as the headline in his own post. His post was brief and written in the same devil-may-care attitude that the company was founded on, and it almost completely defused the situation. He said:

---

13. "Digg.com suffers user revolt," Wikinews, http://bit.ly/cBvFW2 (en.wiki-news.org/wiki/Digg.com_suffers_user_revolt;_Founder_will_not_fight).

14. Kevin Rose, "Digg This: 09-f9-11-02-9d-74-e3-5b-d8-41-56-c5-63-56-88-c0," Digg the Blog, http://blog.digg.com/?p=74.

"But now, after seeing hundreds of stories and reading thousands of comments, you've made it clear. You'd rather see Digg go down fighting than bow down to a bigger company. We hear you, and effective immediately we won't delete stories or comments containing the code and will deal with whatever the consequences might be. If we lose, then what the hell, at least we died trying."

Instead of whining that it was out of their hands, the people at Digg listened to their community members. Digg admitted that it shouldn't have censored information because their community was built on democratic voting that made the most popular stories rise to the top. By muddying the waters with their burying, they'd effectively 'fixed' the election of front page news. They screwed up, but they resolved the situation like pros, quickly and as transparently as humanly possible. The site admitted that it had responded inappropriately. Then it rectified the situation, offering an apology of sorts. Above all, Digg listened to the voice of the community.

You may read that story and think that the decision was a poor one. By allowing that key to be made public, they opened their company up to potential lawsuits, and the fear of a lawsuit is paramount in this ambulance-chasing world. But, consider that Digg would have likely ceased to exist had they not made that decision. Their most vocal members, and a large portion of all of their members, were threatening to leave the site. There were certainly competitors to Digg that would have accepted a huge chunk of their user base with open arms. Likely, the powers that be at Digg remembered a similar backlash at Friendster that paved the way for MySpace, and chose to keep their members at all costs. Not a conventional decision, and not one you'd learn in business school, but one that paid off in spades (and likely helped the founders sleep a bit better at night).

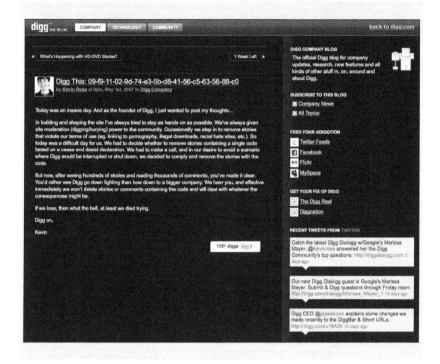

Accountability

This is a big one for any company, but especially for one that boasts a community. When someone makes a mistake and creates a problem or an inconvenience for you, you expect an admission—along with an apology and assurances that the issue is being fixed. Of course, your customers (your online community) expect nothing less from you when you make a mistake. Admit your faults, and then explain how you plan to rectify the problem.

Case Study: Dell

Dell is one of the best examples of true customer accountability. In June 2005, popular blogger Jeff Pulver unleashed a blogosphere backlash at Dell when, after reading his complaint about the Dell customer service experience, he apparently struck a chord with other Dell victims. So many people sang out in agreement that Dell decided to take significant action, rather than brushing off the online community. Michael Dell said:[15]

> "We screwed up, right? These conversations are going to occur whether you like it or not, O.K.? Well, do you want to be part of that or not? My argument is you absolutely do. You can learn from that. You can improve your reaction time. And you can be a better company by listening and being involved in that conversation."

Michael Dell returned to the CEO chair in January 2007. The company took stock, so to speak, and determined that they needed to completely rework their customer service experience. They cut the number of outsourced customer care partners from 14 to 6. They stopped rewarding agents for shorter calls, which they found led to far too many transferred calls. At one point, when Dell's customer care experience earned the name Dell Hell, around seven thousand of the four hundred thousand callers in a week were transferred seven or more times, in an effort to keep call times low. They've dropped a 45% transfer rate down to less than 20%, and hope to get it even lower. Dell is making an effort to get most calls answered with no transfers. Higher-ranking technicians are encouraged to put away the script when necessary and are now empowered to just give customers the answer to fix the problem, rather than heeding some top-down process that may or may not be in the best interests of customers.

---

15. Jeff Jarvis, "Dell Learns to Listen," BusinessWeek, http://bit.ly/ai2dkk (www.businessweek.com/bwdaily/dnflash/content/oct2007 /db20071017_277576.htm).

Dell was accountable to its customers, and listened to the community. The problem has been addressed in a strong way, and Dell continues to strive toward putting the customer service experience back on solid footing. For a company that was really put on the map for their excellent customer attention and value, that had to be a hard lesson to swallow, but they did it with efficiency and grace. As a result, the Dell community is acting in kind. Negative blog posts are down from 49% to 22%. As Dell's blogger, Lionel Menchaca, said, "That change in perception just doesn't happen with a press release."

Focus On the User

Base your community planning on the overall goal that you want to provide a value-filled service to your user. Anything less wastes your time. Events, contests, giveaways and other cool stuff should be completely designed to help community members, even if there is no real value to you, other than PR goodwill. A site that predominantly caters to women could encourage women to get a mammogram by offering a Pro membership to every woman who participates. Another similar site might have a 'weight-off challenge' around the first of the year to help people be cognizant of their holiday weight gain. It's not all about you. A great way to show your users that you actually do care about them is to do something that provides your company with very little visible and tangible value, but gives users enormous value. That's what we call the 'value proposition.'

Case Study: MyBlogLog

Robyn says: When I was at MyBlogLog, we discussed a contest that would serve the Problogger segment of our community. From the beginning we knew we could massively ramp up our user base by asking people to invite others and rewarding the person with the most users. But that meant only one person in the community—the winner—would benefit. We also knew that we would see a great deal of fake accounts if we ran a contest like this and fake accounts are useless in a real community; beyond increasing our numbers, our number one goal was strengthening our community. Since we really wanted to do something cool and fun for our members, and not just a contest that would benefit only us in the long run, we decided to focus this contest completely on the benefit to our users.

MyBlogLog is a place where anyone can build their own special community, so we challenged them to grow their community by offering a bunch of cool, Problogger-themed prizes to the person whose community grew the most (by percentage of overall increase) during the period of the contest. This way, even those who lost, if they participated, came away with something pretty cool: a larger community.

It did turn out to benefit us, though. We had something to talk about, and all marketers love that. In the weeks that the contest ran, it was continuously discussed in the blogosphere. The buzz from that contest was very valuable. We also saw our site signups increase during the period, because so many people invited their current readers to sign up for our service. But, most of all, we learned from our already-loyal members that this contest was something they enjoyed. Our community members asked us to organize more contests of that nature. We focused on our members and they gave us much love in return.

Simplicity

Make everything you do, from web design to product development, super simple. Even though your customers are not morons, you should design things that even a moron *could* figure out. People don't want to waste their valuable time by coming to your site and struggling through your design mistakes. Robyn's 9-year-old daughter can design a multimedia presentation using a Mac. Apple is a company that understands usability.

Anything community driven needs to be simple. If you put up a message board, make sure that existing community logins tie in easily to the message board. Create simple solutions and designs that allow members of the community to interact easily with other members. Amazon.com provides a simple 'rate it' system. All you have to do is click the stars. The same is true of Netflix. The act of rating doesn't take you to another page, wasting valuable time and discouraging ratings. It's one click and you've made your point. The design is simple, and community members can see what others like. Dell's shopping experience allows you to see the star rating on each of the items in their online store, so that you can take the opinion of others who already own the product into account, when making your computer purchase decision. The idea is very transparent and must give customers a sense of trust in the product—and in their own decisions.

Simple page design, uncomplicated controls and obvious next steps will ensure that your site gets used when people land on it. If you are wondering how simple your own site is, there are three easy ways to find out:

1. Put a computer-challenged family member on your website, and observe as s/he navigates it.
2. Take a look at your server logs to see how long people spend on your site. If it's very short, it's either because your product/service is absolutely worthless or because users couldn't or didn't want to navigate through the mess of your site.
3. Pay attention to your metrics. If everyone gives up on step three of setting up a new account, then there is something preventing valuable conversions. You need to figure out what that is and fix it immediately.
4. Hire a usability consultant and ask them what they would do differently. I've never met one that wasn't unabashedly honest about your failure at usability, so if there's anything wrong, a good UED (User Experience Design) consultant will find it.

Do Unto Others...

The bottom line is to treat the members of your community the way you want to be treated. Do you want to be answered quickly? Should your feelings about a given topic matter to the owner of any of your current offline communities? Would you participate if you couldn't figure out the rules, or couldn't find the building in which you are supposed to meet? Do you want to be recognized for your efforts in the community? Do you want to spend time learning to use an ungainly site? Give your customers the same things you'd want and they'll reward you with loyalty and free word of mouth marketing.

6 Social Media

So far, we have looked at a variety tools that you can use to build community online, and social media is really the discipline of using social sites to do just that. To a certain extent, many of the tips given before apply to social media—and some of them even overlap. After all, leaving comments on someone else's blog is certainly a social use of media. But there are some very specific social media tools that you can use to become more involved in an online community.

General Overview of the Social Media Landscape

Social media is one of the fastest growing segments of online interaction, and marketers are only just figuring out how to use it. Additionally, they are struggling to put a monetary value on its implementation. In April 2008, Andrew Baron, Founder of Rocketboom, put his Twitter account on eBay.[16] When the auction was taken

---

16. Daniel Terdiman, "Rocketboom creator selling Twitter account on eBay," CNET News, http://news.cnet.com/8301-13772_3-9917670-52.html.

down, the bid for his account was sitting at more than $1,000. Microsoft thought Facebook warranted a $240 million investment.[17] Rupert Murdoch made a $580 million cash payment for MySpace.[18]

The fact that none of these companies has managed to figure out real monetization seems to be a non-issue. What does seem to be of issue is that social media companies are being recognized as having value, even if that value cannot be quantified with exact dollar figures. Businesses that want to build online communities should understand that these companies, and social media sites in general, are growing in importance, and they should endeavor to learn more about these emerging tools for PR and Marketing.

General Social Networking Websites

There are general social networking websites that appeal to a wide variety of people. These are sites like Facebook and MySpace, as well as the somewhat specialized, but still widely popular, YouTube and Flickr. Anyone can join and set up a profile, and it's easy, almost seamless, to share content that others will enjoy. Many businesses have started setting up profiles on social networking websites in order to have a more conspicuous presence online, and to humanize their images. If you set up an interesting and active social networking profile, it is possible to build a community that is interested in what you have to say—assuming it's not just mindless PR drivel.

Specialized Social Networking Websites

In addition to general social networking websites, there are specialized ways to network with others. LinkedIn is a professional social networking website. GoodReads is a social networking website for book lovers. ActiveRain is designed for real estate professionals. BlogHer is a social network for women who blog—or male bloggers who blog about women's issues. FourSquare and GoWalla are location-based sites

---

17. Jay Greene, "Microsoft and Facebook Hook Up," BusinessWeek, http://bit.ly/bTBKq9 (www.businessweek.com/technology/content/oct2007/tc20071024_654439.htm).
18. Jeremy Scot-Joynt, "What Myspace Means to Murdoch," http://news.bbc.co.uk/2/hi/business/4697671.stm.

that allow you to check-in at local establishments, and possibly get coupons or tips on what to order. They're like Yelp, but more dedicated to mobile users. If you do your research, you can most likely find a social networking platform that caters to your niche. And if you can't, it is possible to create something similar on another site, like Facebook, or even on old-school sites like Yahoo! Groups or Google Groups.

Social Bookmarking Websites

Did you read something interesting that you want to share with others? Submit it to Digg, tag it in Del.icio.us, Buzz it up on Yahoo! or Google, share it on Facebook, StumbleUpon or Twitter. This type of social media allows you to share interesting information with others, as well as tag things for your own benefit.

Microblogging websites

just found out Umbra was a brand name. i just thought it was the most ass-kicking sword in Oblivion. Learn something new every day :)

about 22 hours ago from Nambu

duzins
Robyn Tippins

One of the biggest trends in social media is microblogging. This is when you share small snippets of your life. You generally get a maximum of 140 characters to share a story, tip or just let people know what is going on right now. Twitter started the microblogging trend, and many brands, like Starbucks, are already making good use of it.

Carefully analyze whom you are trying to reach, and become active in the social medium that is most closely related to your company's audience.

There are several schools of thought on how to use this technology, some tending to be very personable and others playing it much more professional. Zappos (twitter.com/zappos) is a good company to watch on this, as they've really struck a personal, fun tone as a brand. Comcast (twitter.com/comcastcares), on the other hand, has had success with a professional, but caring, voice.

Using Social Media to Your Advantage

Before using social media, it is important to think about *why* you are using it, and what you hope to accomplish. Just signing up and creating profiles is not enough. You need to have a purpose and a plan—and you need to be able to devote the time necessary to creating a presence that goes beyond tactics. Your social media presence needs to be in line with the rest of your online community-building efforts: sincere efforts to serve your community members better.

Notifications

One of the best things about social media is that it is extremely easy to share things with your contacts. Whether you use Twitter, Facebook, Foursquare, or LinkedIn, social media can help you easily reach out to your community members. Find followers on Twitter, and when you have a new promotion, you can 'tweet' it, and your online community members will know about it. You can 'share' information through messages on Facebook, and you can post the latest on your MySpace profile.

There are also opportunities to make your own site more visible, by integrating the Facebook Newsfeed, the Twitter stream, the Yahoo! Updates API, and other similar 'broadcast' web services. That means that each time a user on your site participates in an 'action,' it can be broadcast on Facebook, Twitter, Yahoo! or whatever site you've integrated with, with a clear link back to your site. This increases your reach exponentially, putting your site in front of that site's network of users. Using social media to get the word out is fast and effective.

Sharing

Social media also makes it easy to share. Did you write a great how-to post on your business blog? 'Vote' on it in Reddit, or use your Digg account to share it with Twitter followers. Use StumbleUpon to reach out to people who are interested in finding new websites and companies. You can also use the ShareThis button to put your information on a variety of social media sites, in one quick click.

Case Study: Professional and Personal Connections

Miranda says: It is possible to make connections and 'meet' people that can benefit you in professional and personal capacities. Both LinkedIn and Twitter have allowed me to find new writing opportunities. LinkedIn has resulted in several writing gigs ranging for producing content for websites to freelance journalist positions. When someone is looking for a professional, the first place he or she is likely to look is among those they know. Or to get recommendations from those they know. LinkedIn facilitates this process.

Twitter actually allowed me to find some great guest posting opportunities for my blogs. This allowed me to reach new readers, while at the same time developing relationships with other bloggers.

Growing your connections also means that you can broaden your reach to 'friends of friends' and reach even more people. Plus, it grows the number of contacts you can promote valuable information to. But beyond that, it is also possible to find those with whom you resonate on a more personal level. I have 'met' people through Facebook, Plurk and Twitter that I am excited to meet in person when I go to conferences.

Basic Dos and Don'ts of Social Media

When you engage in social media as an exercise in building online community, there are a few things to keep in mind:

Do: Create a profile that is consistent with your business and your branding. Your social media profiles should reflect the image you want.

Do: Have the same name for each of your social media profiles. You can tweak the profile to match the site, but someone who 'friends' your company on Facebook should be able to recognize you on Flickr or Twitter.

Do: Link to your business website from your profile.

Do: Visit other profiles and 'follow,' 'add as a friend,' or otherwise connect with people who might be interested in your brand.

Do: Offer valuable, useful and/or entertaining content. Make sure that your content is consistent with your image, and be sure that it stimulates dialogue.

Do: Be careful of the pictures and videos you post. Make certain that they are interesting, but also that they are in good taste that is consistent with your brand.

Don't: Forget to update. Your online community will trickle away if your social media profiles and postings are static. Add photos and videos, as well as new content, regularly.

Don't: Become involved in constantly dissing your competitors. Instead, keep things positive—and focused on the good things in your company and your industry.

Don't: Send spam. Many overzealous marketers send a 'shout' every time something new is posted. On Twitter, there are those that do nothing beyond 'tweet' advertisements. This annoys most people. Make sure that you provide content that has a use beyond self-promotion or you'll lose eyeballs.

Don't: Expect immediate results. Realize that building an online community—and using social media to do it—is a process. It takes time, a serious amount of effort, an awesome idea/product, and, honestly, a lot of luck. You aren't going to have an amazing and miraculous explosion of traffic overnight; you may not ever get it. But the idea is to build a solid base of loyal community members who might, in turn, share you with *their* social networks.

Don't: Sign up for *everything*. You would never have time to keep up with it all. Evaluate which social media sites would be most beneficial to helping you build and maintain an online community, and focus on being involved in those.

Don't: Ignore others' attempts to get social media love. If one of your online community members needs a Stumble or a Digg, oblige—just as you expect your community members to help you out.

Some things to keep in mind as you develop a social media strategy

Ask yourself the following questions when you set out to create a social media strategy as part of your efforts to build online community:

1. **What are your goals?** Define your goals for the social media campaign. Do you want to improve your image? Are you looking to get the word out about your company or a special promotion? Do you want to boost traffic? Figure out what you hope to accomplish, and then develop reasonable measures of success to grade your efforts.

2. **To whom are you talking?** Which stakeholders do you want to reach? Define each group: customers, industry insiders, power users, influencers, professionals, etc. Also, consider the smaller groups within the larger. Does your audience include moms...children...singles...religious folks? Will you be aiming your message at a specific segment of the population? You will need to develop your approach based on whom you are trying to reach. You need to go where they are; get involved in the social media site(s) that your audience is using.

3. **Who will be running your social media strategy?** It is important to figure out who will be in charge of updating profiles, becoming involved in discussions and interacting with community members. This is very important because you want a continuity of your image and message. Choose a small number of savvy people to maintain your social media presence. Remember, this is not a role for an intern or someone so junior that they cannot follow your brand guidelines.

4. **What are the rules?** Companies need to develop guidelines for what will be discussed online—and what images are appropriate. Create some reasonable rules for the social media team to follow. But make sure that they have the freedom to do what needs to be done, without constantly running to the higher-ups for approval. It is a delicate balance, but common guidelines (such as no swearing, or deleting spam comments) can help provide direction and consistency in social media.

7 Things to Avoid When Building Online Community

Up to this point, we've focused mainly on things you can do to help you build online community. But what about the things you should avoid? Building online community requires that you care about your community members. You need to show that you are concerned, and adhere to practices that confirm to the members your interest in what is useful to them. Once your community decides that you are concerned only with how much you can get, your members are likely to fade away, shunning you—and encouraging others to reject you as well. There are some definite practices that can get you shunned online, and once your online community shuns you, it is difficult to become 'un-shunned.' The best thing to do is to avoid these common practices that make you look as though all you care about are page views, subscribers and click-throughs.

Insincerity

This is a big one. You *do not* want to come across as insincere. Do not say or do things that you do not mean. When you apologize, make sure that you mean it, and that you try your best to make amends. Eventually, lies will be caught out, and self-serving actions will be discovered. Examine your motives and do your best to create a culture of caring within your business, so that sincerity becomes a part of your online operations.

Sincerity and consistency in your personality and brand image are also important. Do not try to be someone you are not. Instead, make sure that you develop your personality consistently. You want your personality and brand image to be consistent across online locations. There are a number of forums, social media sites and other Internet places for you to interact with community members. Increasingly, the Internet is becoming a smaller place. It is quite likely that you will 'run into' community members on forums and view their profiles at online locations beyond your own site. When your community members encounter you elsewhere online, you want them to recognize you as the person they know and like from your website. This is very important. If you want to develop a good online presence, then you need to be yourself everywhere you go.

Your online personality and image should match your offline personality as well. If you own a business, chances are that you need to work with community members and potential business partners offline. People who 'know' you online should not be surprised when you meet offline. Your personality and brand image should be consistent. This is why it is so important to be yourself. Trying to keep up a charade all the time can be tiring—and eventually someone will see through your insincerity and broadcast it.

Impersonal

Because the Internet is a place full of online communities, your community needs to be personal. Your community members want to feel as though you are providing products, services or online experiences that fit their individual needs. If you are impersonal, community members will leave (assuming they even join in the first place). Community members want to be able to customize their experience to a certain degree. Wish lists, saved settings and even the way the text looks when they engage in chat attract members and help them feel as though what you provide is truly 'theirs.'

It's not that hard to develop programs that can insert community members' names into newsletters and other communications. How do you feel when you get snail mail addressed to *resident*? Or when a letter is headed 'Dear Friend?' Community members feel that same reaction when they receive emails and other messages addressed similarly. Do you want to be referred to as 'Community Member' or 'Valued Customer?' Of course not! And your community members do not want to be, either.

Other things you can do (especially retail websites) to personalize the experience for online community members is to add special touches. Have lists of favorites. Allow members of your community to share their lists and preferences with others. Auto-suggest items they might like based on what they've purchased, viewed or read before. Consider letting members choose different colors as backgrounds on their profiles and account pages. Some services allow users to design their own personalized backgrounds by uploading their own images. Find a way to let your community members make a small corner of your website their own. If they've put down roots in an area, they are more likely to return again and again.

Lying

DO NOT LIE, EVER. Someone, somewhere, sometime will call you on it. With thousands (or even millions) of people visiting your website or combing through other things you have done online (and even offline), eventually lies will be found out. Do not ever purposely mislead your community members. This will destroy your credibility and prompt a mass exodus from your community.

If something *does* go seriously wrong, do not try to cover it up. Do not dance around the issue. If you try to avoid it, many in your online community will see that as a lie. Instead, confront problems and issues head on. Apologize, if it is your fault. Do not blame others; take respon-sibility. Then let everyone know how you plan to fix the problem. Follow through, correcting the mistake and making recompense. Then do your best to move on.

Honesty is always better received than shiftiness. Most people are willing to offer you a second chance if you are forthright and make a true effort to improve the situation.

Blatantly Ask Others for Links and Mentions

Watch out for blatantly asking someone to link to your work or your website. It's one thing to ask someone you know reasonably well to Stumble or Digg something that you've written, once or twice a year. If you have done something of good quality, and you think that it has true use for others, you can ask someone you know fairly well in your online community to submit it on your behalf (but you'd better be willing to submit something of theirs in return, sometime down the road). It's another thing altogether to email a blogger or another website owner and say, 'Please link to me!' It's rude. Asking for 'link exchanges' as part of a 'directory' or with the intention of trying to boost your search engine optimization without adding truly useful content is another no-no. It's just gross and screams icky.

If you want people to link to you, or to mention you, you need quality content that is useful and interesting (or at least entertaining). You also need to be a good member of the online community yourself. Instead of constantly begging others for mentions, link to other bloggers' work. Most bloggers and website owners will notice your efforts and, if you have a quality community and a good website, reciprocate. Leave relevant and useful comments on the websites of others in your niche.

When you do link to something or mention something someone else did, you can email that person and let him or her know. For example:

"I really enjoyed your post on subject <x>. I mentioned it on my blog today: [paste in the web address here]. I think that you made a really valid point on <this subject>, although I disagree with you slightly on <this other thing>. I look forward to reading more on <this subject> on your website in the future."

Notice that this message doesn't include a plea for the person to recip-rocate. It simply expresses appreciation, shares a link to your post and offers a couple of constructive thoughts. More than likely, the other person will be curious and check out your website, leaving a comment on your post. Perhaps he or she will even promote the fact that you mentioned the blog or website in a post he or she writes later. But you should never ask for that kind of promotion, and you can't expect it either. Some people aren't going to link to you, no matter how interesting you are, and others will link to you often. It's very similar to offline word of mouth. You can do a few things to get noticed initially, and if your product is good then word of mouth will blossom. However, like real life, if you go too far and push too hard, you will come off as smarmy, alienating the very people you hope to impress.

Most websites and blogs have places where you can leave your web address when you make comments. Be sure to fill in the field, but don't go further than that. No more than one link, period. And, again, make sure that your comments are useful and engaging. That will provide a way for people who were intrigued by your comment to visit you if they so choose.

Case Study: HabitatUK

A boutique furniture store, HabitatUK, recently saw some shameful PR due to their misuse of Twitter.[19] For a little backstory, you need to know how Twitter's hashtags work. If you are familiar with tags—which is basically what a hashtag is—you know that tags are a way to classify information—in this case, a tweet. When someone tweets, and they add the text #iphone, then you know that their tweet is about the iPhone. If someone tweets and adds the text #recall, the story may be alerting you to a food recall.

This is important because those hashtags are used to find information on those subjects. At search.twitter.com, you can search for #iphone, and see real-time tweets about the iPhone. But, more importantly, in the sidebar, there is a list of 'trending topics' that show the hottest searches and hottest hashtagged items across the entire site (i.e. the world). The Iranian revolution was a trending topic for weeks, tech conferences usually are a trending topic for at least a part of the conference, and many products get listed as trending

---

19. Tiphereth, "How not to use Twitter: HabitatUK as a case study," Social Media Today, http://socialmediatoday.com/SMC/103334.

topics: the Kindle the day it came out and the iPhone each time Apple released a new one. When the iPad came out, it trended for over a week.

Spammers, usually sleazy marketers and 'ladies of the evening' offering up webcam services are using those trending topics to advertise their wares. During BigOmaha, a conference Robyn attended in 2009, close to 20% of tweets that used the #BigOmaha hashtag were spammers, just wanting the attention of those following the conference.

Apparently, HabitatUK decided to try a similar tactic. Here are some screenshots Robyn took from Google's cache of the HabitatUK Twitter Account (@HabitatUK[20] on Twitter).

#MOUSAVI Join the database for free to win a £1000 gift card http://bit.ly/2wPLO ? Now!!

7:57 AM Jun 15th from web

#TRUE BLOOD Join the database for free to win a £1000 gift card http://bit.ly/2wPLO ? Now!!

7:57 AM Jun 15th from web

#KOBE Join the database for free to win a £1000 gift card http://bit.ly/2wPLO ? Now!!

7:57 AM Jun 15th from web

#AT&T Join the database for free to win a £1000 gift card http://bit.ly/2wPLO ? Now!!

7:57 AM Jun 15th from web

As you can see, HabitatUK used the hashtags for AT&T and Mousavi, who was hot at the time, due to his contesting of his defeat to Iranian President Mahmoud Ahmadinejad in the disputed 2009 election.

---

20. http://www.twitter.com/HabitatUK

We'll let you decide if this is spamming, but we can let you know what a few of their followers thought of it by looking at their tweets:

@mattfarrugia (Matt Farrugia) – It's hard not to label @HabitatUK as a spam-bot. Terrible thing to do to a premium brand.[21]

@GrrAargh (Phil Waters) – @HabitatUK Spamming news of important events. You must be so proud.[22]

@drewm (Drew McLellan) – Wow, @HabitatUK really need to clean up their act. Not what you'd expect from an otherwise classy brand (via @roshorner).[23]

Habitat did, at least, admit their error:

"This was a mistake and it is important to us that we always listen, take on board observations and welcome constructive criticism. We will do our utmost to ensure any mistakes are never repeated."[24]

What's important here, is that they didn't understand how the service was used before they began using it. It's a common mistake that brands, large and small, make. They jump in, thinking any idiot could get this right, and they blow it by doing what they've always done. Marketing has always been about getting more attention for the product you're selling, and that's exactly what Habitat was doing. However, social media marketing is really quite evolved. It's much more of a conversation than a sales pitch, and that's why so many brands hire consultants who 'get it' to carry out their social marketing.

---

21. http://www.twitter.com/mattfarrugia
22. http://www.twitter.com/GrrAargh
23. http://www.twitter.com/drewm
24. http://bit.ly/cgm9V8 (www.telegraph.co.uk/technology /twitter/5621970/Habitat-apologises-for-Twitter-hashtag-spam.html)

Spam

The subject of spam is closely related to soliciting links. Spam comments are generic and blatant attempts to self-promote. They rarely contribute anything useful to the discussion at the website. No one likes spam comments. Avoid leaving comments on others' sites, forums and blogs that say things like, 'Come check this post on _____ at my website!' Likewise, simply filling in the fields and then saying 'Great post!' is frowned upon. What have you added to the conversation? Answer: Nothing.

Many web masters, forum administrators and bloggers remove comments that are outright spam—and some that even resemble spam. It does you no good to leave worthless comments that are only removed. Forums, blogs and websites that actively remove spam know that their own credibility diminishes, if they leave spam entries up. If you continually spam, some will ban you from their discussions.

You should leave thoughtful and insightful comments. Try to add to the conversation. Instead of being viewed as an annoyance, you will be deemed a contributing member of the online community. This is an excellent way to share your expertise and gain new members for your own community.

Poorly Researched PR Pitches

One of the biggest pet peeves that other bloggers and online news media outlets have are PR pitches and press releases that are irrelevant or un-researched. Before you send a press release or interview pitch, make sure that you are suggesting something that is relevant to the site or blog.

Case Study: Mommybloggers

In early 2008, the blogosphere erupted in a backlash against PR people who were trying to get mommybloggers to endorse their items. Mommybloggers is an influential group, and in many cases PR folks try to fit these diverse women into a neatly labeled demographic. Bloggers (who also happen to be moms) were being invited to try products that did not relate to their subject material, or being asked to promote products they don't use on their sites. Unfortunately, they were either being pitched *only* as mothers, or their needs as mothers weren't being addressed.

One of the biggest issues was to do with Johnson & Johnson's Camp Baby. Influential mommybloggers were invited to an all-expenses paid event that would allow them to try products and have a good time. The idea was to give mommybloggers a great experience involving a number of products and services. The hope was that buzz for Johnson & Johnson would come from legions of mommybloggers raving about the products on their blogs.

What ended up happening was poor execution.

One mommyblogger was disinvited after she wanted to bring her breastfeeding infant.[25] Another was disinvited because she couldn't stay the whole time.[26] Johnson & Johnson didn't have accommodations for breastfeeding infants (or any children, for that matter), and the company made it clear that they didn't want to 'waste' the trip on someone who couldn't stay the entire weekend. Obviously, it is poor form to invite someone to an event and then withdraw the invitation when your guest doesn't meet the exact the parameters that you prefer. Mommybloggers erupted in righteous anger and began talking—and blogging—about how they should be treated as *people* and not just numbers.

---

25. "No babies allowed at Camp Baby," mothergoosemouse, http://bit.ly/cecP6B (mothergoosemouse.com/2008/03/18 /no-babies-allowed-at-camp-baby/).
26. "Because I needed something to talk about at BlogHer Business," CityMama, http://citymama.typepad.com/citymama/2008/03/because-i-neede.html.

Many PR professionals are simply clueless when it comes to social media and end up with public humiliation when they do not heed repeated warnings from bloggers. We've seen many people called out by name on popular blogs because they finally annoyed the blogger with one too many emails. Do your research and don't pitch your company, product, blog or site to a blogger who isn't even covering news related to your industry.

Another thing to remember is to keep story ideas and press releases short. In as few words as possible, explain why the blogger or site owner should care—and illustrate why readers would be interested. Take a little time to figure out what the blog or site is *about*. Don't waste their time and please don't say you read their blog if you don't (lying again). Remember, most full-time bloggers and online journalists get at least a few of these pitches per day. Most go in the big, round filing cabinet, but a few lucky ones are chosen because the pitch would make a great story for their readers. And, a few unlucky few will be given the public humiliation treatment. Take care to avoid that outcome at all costs.

You would never pitch a writer at the Washington Post or Vogue without understanding what his or her community members want. While bloggers aren't as powerful as these media outlets on their own, when consensus is reached, they can be even more powerful. Besides, as magazines and newspapers begin to fade away, blogs will become more powerful. Remember, what is written online does not disappear. It's there as long as the site is up, and search engines will treat bloggers' tirades against your company with sufficient weight that you may greatly regret annoying them. Remember 'Killer Coke?' What people say about you online can be really great, or really terrible, and either way it's there to stay. When pitching your brand to bloggers in an effort to attract more community members, make sure you take a little time to visit the site and learn what it's all about.

This is also a valuable lesson for interactions with your community members. Understand what they want, and what makes them tick. If you want to build your online community, and do it effectively, you need to pay attention to how you treat your community members.

In Closing...

Now that you know that online communities are here to stay, and that they can be as vital to your personal well-being as to the success of your business enterprise, it's time you started your own online community!

We hope we've shared with you some of the critical distinctions of participation and functioning in online communities. Online, your voice soars far beyond your immediate neighborhood, making it especially important that you project a clear and consistent message and personality. And, because the online world is *sticky*, your words, thoughts and feelings will reside online long after they've been erased from your own memory. Staying aware of these differences between the online and offline worlds will help you communicate online with thoughtfulness and confidence.

Finally, to thrive in your online community or communities, remember what you've known since kindergarten: share, play nice and never run with scissors.

About the Author

Robyn Tippins is a community advocate with over 10 years experience in the social media space. From her early days marketing her own small business using forums and email lists, to blogging, podcasting, vlogging and video game immersion, she's often used social networking to engage and communicate. In her current role, Robyn oversees the community aspect of the external developers on the Yahoo! Developer Network.

Robyn has blogged for blog networks and corporations, podcasted for small and large businesses, worked closely with social networking sites, and advised Fortune 500 companies on social media and community. Her early podcasts featured some of the web's most interesting and well-known Web 2.0 experts in fields such as VoIP, Technology, Open Source, Marketing, Social Networking, Video Games and Blogging.

She finds her greatest joys in moments away from her computer, spending time with her husband and four children, ages 4 to 11. She and her family reside in the San Francisco Bay Area of Northern California.

Miranda Marquit is a professional blogger and freelance writer working from home. She has five years experience in the blogging and social media space, mainly providing content and support for corporate blogs. Miranda understands the importance of blogging and social media in online marketing and community building, and enjoys interacting and networking via the Internet.

In addition to professional blogging, Miranda is a freelance writer with a Journalism degree. Her work has appeared in national magazines and on news Web sites. She is also a columnist for her local newspaper. Miranda enjoys reading, music, travel, and the outdoors. Her favorite activities involve using her hobbies as a way to spend time with her husband and their six-year-old son. Miranda lives with her family in Logan, Utah.

Other Happy About® Books

Purchase these books at Happy About http://happyabout.info or at other online and physical bookstores.

18 Rules of Community Engagement

This book is a definitive guide for those seeking to facilitate and grow online communities and develop social media strategies for themselves or their organizations.

Paperback: $19.95
eBook: $14.95

42 Rules for Successful Collaboration

Readers of this book will walk away with a much better idea how to be successful in their interactions with others via the computer. It will help people who are on teams separated geographically, as well as managers and executives. The book filled with high-tech nuggets of wisdom for programmers and IT professionals. But it also has practical rules that apply to anyone who works with others.

Paperback: $19.95
eBook: $14.95

Social Media Success!

This book is a launch pad for successful social media engagement. It shows how to identify the right networks, find the influencers, the people you want to talk to and which tools will work the best for you.

Paperback: $19.95
eBook: $14.95

42 Rules of Social Media for Small Business

Written by communications professional Jennifer Jacobson, this book is designed to help working professionals find social media that fits their business and get the most out of their social media presence.

Paperback: $19.95
eBook: $14.95